PERSPECTIVES ON THE TELEPHONE INDUSTRY

PERSPECTIVES ON THE TELEPHONE INDUSTRY:
The Challenge for the Future

Edited by

James H. Alleman, Ph.D.
International Center for
Telecommunications Management
and
Richard D. Emmerson, Ph.D.
INDETEC Corporation

A United States Telephone Association Book

1817

Harper & Row, Publishers, New York
BALLINGER DIVISION

Grand Rapids, Philadelphia, St. Louis, San Francisco
London, Singapore, Sydney, Tokyo

Copyright © 1989 by Ballinger Publishing Company. All rights reserved.
No part of this publication may be reproduced, stored in a retrieval
system, or transmitted in any form or by any means, electronic,
mechanical, photocopy, recording, or otherwise, without the prior
written consent of the publisher.

International Standard Book Number: 0–88730–376–5

Library of Congress Catalog Card Number: 89–31297

Printed in the United States of America

Library of Congress Cataloging-in-Publication Data

Perspectives on the telephone industry: the challenge for the future
  /edited by James H. Alleman and Richard D. Emmerson.
      p.   cm.
  Papers presented at a conference "Local Exchange Pricing: the Chal-
lenge of the Future," organized by the Alternative Pricing Concepts
Committee of USTA (United States Telephone Association) and held
Nov. 16–18, 1987 at the Hyatt Regency in New Orleans, La.
  Includes index.
  ISBN 0–88730–376–5
  1. Telephone—United States—Congresses. 2. Telephone compa-
nies—United States—Congresses. I. Alleman, James H. II. Emmerson,
Richard D. III. Alternative Pricing Concepts Committee of the USTA.
HE8815.P42   1989
384.6′ 0973—dc19   89–31297
                                                                      CIP

89 90 91 92 HC 9 8 7 6 5 4 3 2 1

>> *Contents*

v

Part V   Perspective of the Practitioners: The Cost Allocation Drivers

Part VI   Perspective of the Futurists: The Technology Drivers

≫ List of Figures

>> *List of Tables*

>> Foreword

This book is a compilation of papers presented at a conference that was organized and sponsored by the Alternative Pricing Concepts Committee of USTA (United States Telephone Association). Motivated to establish a coherent view of what is taking place in the local exchange arena, we were prompted to organize a conference that would bring disparate views to a common forum in order to explore the elements of their agreement and, more importantly, their disagreement. The conference, "Local Exchange Pricing: The Challenge of the Future," was held November 16–18, 1987, at the Hyatt Regency in New Orleans, Louisiana.

The conference provided a series of educational and working sessions concerned with pricing policy issues and concepts for regulated telecommunications services. A professional, scientific, multidisciplinary forum was created for telephone companies and regulatory staff, researchers, and practitioners to discuss current issues and to illustrate ways in which individuals can develop their own approaches. The program allowed participants to present procedures and frameworks for handling current and future challenges. As the reader will see, the views expressed in one paper may well be in conflict with those of others. But by bringing the parties together, everyone can judge what must be done to convince their "opponents" of the virtues of their position.

The committee is pleased to bring these issues to a wider audience. The committee's hope is that the reader will be able to rethink the options and the policy alternatives available while there is still time to effect meaningful change. The stakes are too high and the outcome too important for the issues to be obscured in regulatory, legislative, and judicial processes. The viability of local exchange carriers is at stake.

Before closing this foreword, I would like to acknowledge some of the many people who made this book possible: the authors, who contributed to the conference and to this volume—they clearly have set the terms of the debate. USTA, which sponsored the conference and, in particular, Stephen Burnett, who as a staff director of USTA supports the committee and helped to see the book through its legal requirements. Tina Knight and Donna Haegele, on his staff, supplied support at critical junctions.

In addition, I would like to thank the committee members who devoted a significant amount of time to creating the conference, and now this book. They include: Wendy Blumling, Southern New England Telephone; Neill Dimmick, Kozoman and Kermode; Cathy Dudderar and Lyle Roberts, Centel Corporation; Thomas Gorman, Yelm Telephone Company; Warren Hannah, United Telecommunication, Inc.; David Hostetter, Southwestern Bell; Richard Reinking, Mountain Bell; Gene South, Panhandle Telephone Cooperative, Inc.; and Rosemary Spell, Bell Atlantic Corporation.

Finally, this book would not have been possible without the concentrated effort of our editors, James Alleman and Richard Emmerson. They contributed to the high quality of the conference and, as the reader can judge, the superb quality of this book. It would not have been possible without their effort.

—Gerald Cohen, GTE Service Corporation
Chairman
USTA Alternative Pricing Concepts Committee

» *Acknowledgments*

In any publication of this magnitude, many people have to be thanked for their contribution. This book of readings is no exception; in fact, it has had the participation of far more people than many other publications of its size. This brief note cannot pay adequate tribute to those who made it possible, but it will at least acknowledge those to whom we can but give inadequate recognition.

The authors of the following articles are the first to be acknowledged. They endured the many drafts sent to them by the editors, as well as other nonsense. They responded courteously and promptly.

The individuals involved in compiling the book and in text editing and general administration deserve thanks. Sandy Fedor kept one of the editors on track, and Carmella Lanza-Weil did the same for the other editor. And a special thanks is due to Richard Sarro, who in addition to providing text editing and general editorial assistance, was instrumental in proofing the final changes and seeing the book all the way through to completion.

Recognition should be made of the sponsorship of the major telephone companies: Ameritech, Bell Atlantic, GTE, NYNEX,

Pacific Bell, Southern Bell, Southwestern Bell, United Telecommunications, and U.S. West Communications. We are particularly grateful to two individuals at GTE: Leland Schmidt, who, in addition to contributing an article, helped immensely in seeing that the project obtained support, and Gerald Cohen, who allowed the necessary resources to be directed to the project.

The Alternative Pricing Concepts Committee of the United States Telephone Association (USTA) and the USTA have to be thanked for pulling together the conference from which these papers originated and encouraging the production of this book.

Finally, Ms. Carolyn Casagrande at Ballinger Publishing Company not only had the patience to endure the various delays but provided useful guidance and encouragement when it was most needed. She guided the process through from beginning to final result.

JHA
RDE

>> Introduction

This book is a dialogue among some of the major players in telecommunications today. While the papers included in this book came from a conference on pricing, the range of topics covered in them goes far beyond this concept. Evident in the papers are the tensions and contentions among the various stakeholders: the potential winners and losers in a fast-changing and uncertain environment. The themes of the papers meander through technology, regulation, domestic and international competition, to arrive at alternative conclusions about the future of telecommunications.

This set of readings comes at a propitious time: a time when the industry is being shaken by international competition; a time of confusion for telephone companies and their customers; a time when technology, regulation, and markets are changing at a pace unprecedented in the history of telecommunications. The reader will not extract from this book a clear understanding of the path the industry is taking, nor of the point at which the industry may arrive. Rather, the reader will understand the complex issues that must be resolved and addressed in order for the industry to progress. The resolution of the issues addressed here will determine whether the telecommunications industry becomes a powerful force behind that undervalued asset called information, or a passive transporter of bits of data among a few fragmented points. Equally important are the implications for the consumer, domestic and international businesses, professional regulators, and the shareholders of the telephone companies.

The order of presentation is designed to tell a story in itself. The early papers put the current telecommunications issues in a broad global context and add some dimensions to the issues that are rarely considered, even by those working closely within the industry. The papers in Parts 2 and 3 offer competing perspectives from those playing the high-stakes games taking place in the regulatory and market arenas. Parts 4 and 5 cover some of the economic and technological constraints and opportunities facing everyone connected with the telecommunications industry today. Part 6 begins to look ahead, searching for insights to future possibilities that may arise out of the current chaos. Although each reader will enter and exit the readings with different perspectives and projections, one cannot help being dramatically influenced by competing perspectives. We, the editors, will wager that the readers exiting this collection of papers will do so with greater agreement and consistency of perspective than when they entered.

—James H. Alleman
—Richard D. Emmerson

PART I

The Telephone Industry: The New Economic Environment

1

» *The Future of the Public Network: From the Star to Matrix*

Eli M. Noam

» *Introduction*

It takes a better idea to deal with an outmoded one. Monopoly was one concept, deregulation is another. There is no reason to believe either that deregulation is the last word or that there will be a simple swing back of the pendulum. Regulatory concern in telecommunications has paralleled the stages of the industry itself. The monopoly stage of industry structure was accompanied by the regulatory stage of price and profit regulation. The breach of monopoly was tracked and sometimes facilitated by regulation focusing on industry structure. And now we are reaching the next stage, in which the network is rearranged from a centralized, starlike structure into a matrix of interconnected but decentralized networks. This moves the focus of regulation to encompass (besides traditional consumer protection) networks protection—mediating, where necessary, the interaction of the various carriers, network operators, users, and equipment manufacturers.

There are technical, legal, economical, political, and international aspects to this role of compatibility regulation. And its formulation includes dimensions that traditional state regulation did not require in the past. This is one reason why state regulation has, for twenty years, been a follower rather than a leader. It is time, therefore, to think about where telecommunications is going, and to set a regulatory agenda based on the future rather than on the past.

» *Origins*

Because several of the major changes in telecommunications policy originated in the United States under a conservative political regime, they are often viewed as the product of particularly American business interests, wrapped in a Chicago economic ideology. But more recently, several other industrialized countries have begun to adopt similar policies, or at least to discuss changes that previously seemed unthinkable. This raises the question of whether the changes go deeper than the nature of the respective governments in power, and whether they reflect a more fundamental change. The policy changes are indeed part of a broad transition in which the traditional notion of the public network—centralized, closed, and public-spirited—is evolving into a new idea of a decentralized, open, and private-spirited network. This evolving network resembles a loosely interconnected federation of subnetworks, much like the system prevailing in transportation.

Because the network is a social system, in order to understand its dynamics one must go back to its origins. The origins of the centralized network system preceded electronics and telecommunications by centuries and lie in the emergence of postal monopolies. A key date is 1505, when the Hapsburg Emperor Maximilian I granted exclusive mail-carrying rights to what one would call today a multinational company, the Italian family firm Taxis. This concession proved to be an unexpectedly rich source of revenue to the Hapsburgs, who shared in the profits, but it also required vigilant protection from the incursion of other mail systems, of which there was a multitude (Dallmeir 1977). Neighboring Prussia went one step further and in 1614 established a state-run postal monopoly (Stephan 1859). Thus, the post telephone and telegraph (PTT) system was born as a creation by the absolutist state for the absolutist state. While much later this system would be rationalized as based, depending on one's point of view, on economies of scale, national sovereignty, cross-subsidies, or public infrastructure needs, the early creators of the postal monopoly were quite forthright in their mission to make profits for the state and its sovereign. The system became a major source of revenue just at a time when European rulers had insatiable needs for it. This goose with its golden eggs was to be ardently protected through the centuries against encroachment by private competitors and other states.

When the telegraph emerged in the nineteenth century, it was rapidly integrated into the monopoly system. Later, much was made of the military importance of state control over telegraphy. This may have been important for the major powers, but much less so for smaller countries.

When the telephone made its appearance in 1876, it, too, was soon integrated into the state monopoly once its financial viability became clear. Here, official biographies claim that the purpose was to bring telephony to rural areas neglected by commercial interests. This is sometimes true, but in other instances the historical record is quite different (Holcombe 1911). In Norway, for example, private firms served the countryside while the state system took for itself the more profitable cities. At the same time, telecommunications was also integrated into an international system of collaboration—with the officially proclaimed goal of technical coordination, but also, from the beginning, with a cartel agenda on prices and service conditions (Codding and Rutkowski 1982).

For almost a century, a tightly controlled system of telecommunications was in place in most advanced countries. Its structure was supported by a broad political coalition that can be termed the "postal-industrial complex." It included the government PTT as the network operator and the equipment industry as its supplier, together with residential and rural users, trade unions, the political left, and the newspaper industry, all of whose postal and telegraph rates were heavily subsidized.

The system worked to the particular benefit of the equipment industry. The PTTs, through their huge procurements—especially after World War II—provided large markets for the industry. Even better, these markets were almost totally protected from foreign competition by buy-national policies. Within most advanced countries, domestic equipment manufacturers often collaborated with each other in formal or informal cartels that set prices and allocated shares of the large PTT contracts.

In the United States, the structure of telecommunications, although private, was not all that different from the PTT model: a near monopoly, with an integration of the network operator and equipment manufacturing. Its corporate ideology was shaped by AT&T's patron saint, Theodore Vail, himself a former postal man as the head of the U.S. Railway Mail Service.

» *Privitizing the Public Network*

Despite its public popularity, the centralized model of the public network has been subject to forces of centrifugalism that have undercut its stability. Technology is one of the reasons, though one should not exaggerate its contributions. Similar changes have started to reach the electric utility industry, despite the relatively small amount of change in the underlying technology in that field.

The driving force for the restructuring of telecommunications has been the phenomenal growth of user demand for telecommunications, which in turn is based on the shift toward a service-based economy. The shift in highly developed countries towards such activity was partly due to their loss of competitiveness in traditional mass production vis-à-vis newly industrialized countries. It also was partly due to a large pool of educated people skilled in the handling of information. Information-based services, including headquarters activities, therefore emerged as a major comparative advantage of developed countries. Manufacturing and retailing, at the same time, became far-flung and decentralized.

Electronic information transmission—that is, telecommunications—became increasingly important to the new services sector, as well as a major expense item. This made the purchase of communications capability at advantageous prices more important than in the past. Price, control, security, and reliability became variables requiring organized attention. This in turn led to the emergence of the new breed of private telecommunications managers whose function is to reduce costs for their firms, and who for the first time established sophisticated telecomunications expertise outside the postal-industrial coalition. These managers aggressively sought to establish low-cost transmission and customized equipment systems in the form of private networks of power and scope far beyond those of the past. These private networks, whose operation and administration may require hundreds of skilled technicians and managers, carve out ever larger slices from the public network.

The growth of technological and operational alternatives has undercut the economies of scale and scope once offered by the centralized network. Economic and technological development has led to an increased specialization and to a divergence rather than convergence of options. Application options have increased considerably with technology.

By their very nature and tradition, network operators provided standardized and often nationwide solutions, carefully planned and methodically executed. For the large users dependent on telecommunications, this was not enough. In the old days, sharing a standardized solution was acceptable to users because the consequential loss of choice was limited and outweighed by the benefits of the economies of scale gained. As the significance of telecommunications grew, the costs of nonoptimal standardized solutions began to outweigh the benefits of economies of scale, providing the incentive for nonpublic solutions. Furthermore, some users aggressively employed differentiation of telecommunications services as a business strategy to provide an advantage in their customers' eyes, and therefore affirmatively sought a customized rather than a general communications solution.

Another factor contributing to more specialized telecommunications networks is the growing number of groups in society that interlink via telecommunications. Their communications needs as collectives become more specialized, and private user clusters emerge. Early examples were travel agents and airlines, automobile parts suppliers, and financial institutions. Thus, pluralism of association leads to group communications, located somewhere between private and public network activities. As in a Greek drama, the unity of the centralized network unraveled because it reflected the realities of the past. It still had politics on its side, however, and the support of several of the primary organized constituencies in industrialized countries. But the new interests created their political constellations, too. If the telecommunications system is seen as consisting of four major constituencies—equipment suppliers, network operators, employees, and users—the traditional postal-industrial coalition joined primarily the first three, allied with the small-user part of the fourth. Now another grouping emerged: the alliance of large users, including transnational firms, with the most advanced part of the equipment industry, which consists of the computer, components, and office equipment firms.

In Britain, the new coalition was slower to gather owing to the relative weakness of the advanced electronics industry and a defense by the traditional alliance that was more tenacious and ideological than in the United States. However, once the government withdrew its support from the traditional arrangement and instead blessed the service sector by targeting London as the service-sector capital for all of Europe, the postal-industrial complex had to retreat. A similar story can be told for the Netherlands. In Japan, where the first electronics industry transformed itself better than anywhere else, the changes were smoothest, since the equipment industry did not stand to lose much.

We are merely at the beginning of what will be a lengthy process of change in the network, as centrifugal forces encourage the evolution of a new structure of telecommunications. The main principles of this open network system are described in the following.

» The Network as a Matrix

A Network of Networks

The future network concept is one of great institutional, technical, and legal complexity. It includes national and regional carriers, local exchange companies, specialized service providers, cable television companies, domestic and international satellite carriers, local area networks (LANs) and value-added networks (VANs), private networks, and shared tenant services. The network enviroment will consist of an untidy patchwork of dozens, or even hundreds, of players, serving different geographical regions, customer classes, and service types, with no neat classification or compartmentalization possible. To the tidy minds of traditionalists, this is heresy. The future telecommunications network environment will have carriers engaged in multiple functions, although there will be no shortage of official attempts to establish order.

Substantial Absence of Central Control

The central characteristic of the open network model is lack of central control, with no single entity being in charge of an overall plan for the network. Instead, the network becomes a composite of numerous separate planning decisions, moving from the model of the planned system and more towards an "invisible hand" mechanism. The traditionalist perspective was that of chain of command, long-range planning and integration. "The system is the solution" was AT&T's battle cry. To leave this system to the vagaries of hundreds of uncoordinated and selfish actors seems to invite disaster. Can it work? Perhaps this is not the right way to frame the question. Can there be a stable alternative in economies that otherwise favor a market mechanism and that want to stay on the leading edge of applications? In any event, it is quite likely that the dominant public carriers will exercise market leadership for a long time and thus provide the backbone around which private actors will plan and standardize. In that fashion, some central control will continue.

Unsustainability of Most Regulation

Telecommunications is in the process of changing from one of the most regulated industries to one of the least regulated. There are several reasons for this. One is that the increasing complexity of the system makes it increasingly difficult to structure consistent rules. The American experience with the Federal Communications Commission's (FCC's) computer decisions gives an early hint of this difficulty.

Second, rules are unlikely to be enforceable. The subjects of the regulation—streams of electrons and photons, and patterns of signals that constitute information—are so elusive in physical or even conceptual terms, and at the same time so fast and distance-insensitive, that a regulatory mechanism, to be effective, must be draconian, and for that the traditional system has neither the will nor the political support. This means that, to a significant extent, telecommunications will move out of the realm of the political process.

Public System as Core

The telecommunications system will evolve into a mixed public-private arrangement. The public network will not cease to exist. It is likely to remain the core of the system and its prime standard-setter. It deserves public support for the same reasons that have existed before, but without the exclusivity that characterized it for over a century. This is comparable to the situation prevailing in transportation. A state railroad system exists in most industrialized countries, often subsidized both directly and indirectly, but it is supplemented by a mixture of trucking firms, airlines, barges, passenger automobiles, and small railroads. No one advocates a transportation system that bans all private trucks just because they reduce the scope and revenues of public railroads.

Interconnectivity

Whereas in the traditionalist model standardization was a key element, the new model is characterized by a stress on interconnectivity. The difference is that between *ex ante* and *ex post*. To reach or maintain agreements on standards, except for very broad issues, will become increasingly difficult as the number of interests and participants multiplies. Instead, standard-setters or coalitions will emerge around which other actors will cluster, since incompatible services will not usually be attractive to users. But the system may not be fully convergent. Some parallel series of varying network standards are likely. Fortunately, electronics is flexible; a brisk industry of information and protocol arbitrage from one standard to another will emerge.

A key requirement for an open network system is that it extends the common carrier principle from users to networks. That is, the notion that networks can interconnect with other networks, even if they are competitors, is the key requirement for the functioning of an open system. In both the United States and the United Kingdom, the establishment of interconnection of new networks with the existing and predominant one turned out to be essential. This principle, however, requires clarification of the charges for interconnection, and this is likely to remain a regulatory question for a long time.

Right of Access

While the right of interconnection deals with networks' linkage with each other, the right of access deals with users' ability to reach, if technically possible, to any network they choose. For example, a landlord's building network should not be able to prevent tenants from reaching a carrier of their choice. And any network must be able to reach any tenant.

In the traditional model, the concept of common carrier access to the public network was an essential element. An open system is more complicated in that it includes many private parties who operate on a contractual basis rather than as public utilities. The extent of common carrier access in a future open system is an important policy question to be resolved. While the system is likely to impose similar obligations on many network participants, these will not be as far-reaching as present common carrier obligations of telephone carriers. It could be argued, of course, that in the presence of competition, common carrier principles would be unnecessary. But competition also exists, for example, in the airline and hotel industries, both of which have in the United States common carrier obligations without price regulation. The notion of nondiscrimination, particularly for infrastructure-type services, is strong and makes it unlikely that a system of pure contract would be adopted.

Universal Service

The traditional public network operated with the obligation of universal service; that is, virtually any interested customer had to be served, regardless of location. In the open network system, the question is whether universal service obligations apply to all participants. The answer will be differentiated. For some of the more specialized services, the obligation will not exist. But for "basic" service, it will continue, and the definition of "basic" is likely to expand. The boundary line will probably be an ongoing issue of policy discussion. For the future, one main function of the public network will be to function as the service provider of last resort, under financial arrangements that may involve subsidies by the government and the private carriers.

Internal Subsidies

In an open network system, it is unlikely that the traditional system of internal transfers from one class of users to others can be maintained. But this does not mean the end of such transfers. There is still ample reason and opportunity to subsidize some categories of service or some user classes, just as in the case of railroads. Revenues for that subsidy can be raised and distributed by taxation and budget allocation, the normal way in which redistribution takes place in society. It is incorrect to consider a monopoly to be essential for redistribution. The justification for subsidies is still strong, not only for reasons of general social policy and regional development but because of the positive externalities that additional subscribers give to the other subscribers. These transfers are likely to be accomplished by outright governmental outlays to service providers or individuals, as well as by tax assessments on telecommunications providers who do not themselves offer a social service such as rural service or low-traffic public telephones.

Nevertheless, the extent of the subsidies may become smaller once the subject is open to legislative debate. Subsidies are likely to become targeted increasingly toward the poor.

Quality and Price Differentiation

There will be more choice, but less equity. Whereas in the past all subscribers had a fairly similar quality of telephone service and equipment, an open network system will have much variation, depending on the preferences of customers and their willingness and ability to pay. There will also be much greater differentiation in the cost of communications. Just as two adjoining passengers in an airplane may have paid widely different prices for their tickets, so will telecommunications users pay very different rates. Those with small usage and few alternatives—namely, residential users—may pay more per unit than large users or other more alert consumers. This could reverse the cost relationship in the public network, where business users subsidized residential service.

Excess Capacity

The open system does not minimize resources efficiently. There will be more excess capacity than in a centralized system. There is nothing unusual about this: almost every industry has excess productive capacity. The effect is to spur every supplier into user-oriented action to make itself attractive. In the telecommunications field, with its low marginal costs, competition will cause periodic price instability. One of the functions of future regulation will be to help in moderating the worst effects of price volatility.

Transnationalism

The traditional centralized system was international in the sense of being a collaboration on the level of government organizations. It held together well because of a similarity in views—the values of engineering and bureaucracy—and because of a common interest in protecting the domestic arrangements. For a long time, national telecommunications administrations participated almost joyfully in the international sphere because they could return home with an international agreement that would buttress their domestic positions. But in the age of satellites, internationalism has become a threat because it provides new tools to large users and to new carriers. International communications are the soft underbelly of the domestic service monopoly. In the long run, telecommunications will transcend the territorial concept, and the notion of each country having territorial control over electronic communications will become archaic in the same way that national control over the spoken (and later the written) word became outmoded. It is just a matter of time before foreign carriers begin providing U.S. domestic service.

ISDN and ONA

The two network concepts—centralized and open—are reflected in the two major initiatives of their respective proponents, the integrated services digital network (ISDN) on the one hand, and open network architecture (ONA) on the other. Both are pure expressions of the underlying network philosophies. A similar comparison could be made between the PTT concept of videotext and the distributed system of data bases in the United States.

ISDN has been by far the more prominent strategy. It is, at its most elementary, an integration of voice, data, and telex networks into a unified "super-pipe." ISDN encompasses several subconcepts. As a move to spread the digitalization of the network to the "last mile," it is squarely within the trend of technology. As an upgrading of the networks to a higher transmission rate, it responds to the data communications needs of larger users; for residential users, the need for a higher transmission rate is less clear, except as a way of creating the proverbial egg (the network) for a future chicken (applications). Overall, it is a technologically positive concept.

The third element of ISDN is integration, which is much weaker in its rationale. True, to put together separate communications networks into one super-pipe is more elegant from a technologist's view, but from the user's perspective, the cost, performance, and choice of services are what count. Integration is a standardization process, which is always a trade-off between the cost reduction of streamlining and the benefits of diversity. A process of integration is usually a reduction of options. Users are interested in choice for selection, while network operators may be more interested in providing standardized options.

If elimination of "wasteful duplication," which is almost always asserted rather than quantified, is crucial, then in the final analysis the entire economy should consist of one giant integrated enterprise. Clearly, there are organizational diseconomies to the reduction of duplication. The implicit assumption in the "economies of scope" justification for the super-pipe is that cost functions are static in, for example, telephone and telex networks. Yet economists would expect that two services under rival control would usually result in a dynamic downward shift of the cost curves—owing to the extra efforts of competitors—in contrast with the monopolistic situation of unified services. The effects of these downward shifts in costs can more than outweigh the economies of scope.

Thus, there are practical and theoretical problems with the concept of integration, to which several others can be added (Noam 1987). Why, then, is ISDN pursued so vehemently in most developed countries and in their international organizations? Those holding the hierarchical concept of networks are utterly captivated by ISDN, which reaffirms their view of the network as a centrally planned and exclusive system while providing them with a powerful defense against the centrifugal forces of network fragmentation—as well as with profitable business opportunities.

Strictly speaking, ISDN as a technical concept does not negate the possibility that multiple ISDN networks and "networklets" may coexist, compete, and interconnect. There is no notion of exclusivity inherent in the concept of ISDN technical integration. But attitudinally, absence of exclusivity is almost impossible to accept. After all, the elimination of duplication is the primary rationale for ISDN. To permit multiple integrated networks would defeat the entire purpose. From the PTTs' perspective, ISDN strengthens standard-setting control over data communication, reducing the computer and office equipment industries' ability to do so and extending standard-setting and approval power upstream towards equipment that has previously been in the domain of the computer industry.

While ISDN is the archetype for the centralized model and its dynamics, the matrix model has also moved into its next phase—open network architecture. ONA is the framework for opening the core of the public network, central switching. It disaggregates switching into its component functions and permits separate access, interconnection, substitution, and competition with each of them. It provides greater ease in establishing layers of software-defined networks superimposed on the basic transport functions. Different communications providers want to use different configurations of the building blocks of the central switch, and ONA permits these outside parties the use of the building blocks of their choice and the resale of the new service combinations thus created.

An early example for the basic idea was Centrex service, where the local exchange company, in effect, rented out part of its central switch capacity to serve as the equivalent of a private branch exchange (PBX) and provided competition to PBXs by interconnecting the terminals in an organization. As more complex configurations of exchange functions, equipment, transmission, and software emerge, ONA will permit the creation of flexible service packages. At present, central offices depend on giant switches and on extraordinarily complex software. ONA creates a modular approach and enhances the ability to tailor and modify features. In that sense, ONA is part of a movement towards the network of a distributed rather than a hierarchical architecture. In this, it follows the lead of computing, which also started as a highly centralized operation and moved toward a distributed structure. ONA disaggregation is technically compatible with ISDN's integration, since they operate along different dimensions. But ideologically they are at odds, at least for traditionalists who seek to protect monopoly.

Thus, the two network models are pushed into ever more refined expressions of their proponents' interests. Are these developments stable? The centralized model is challenged by the centrifugal forces described above, while the open network concept is only viable if economies of scale and scope are not of a magnitude that would implode the entire network of networks right back into one giant network. But this seems unlikely; even if many of the new ventures in services collapse, the genie of diversity is out of the bottle. Communications are becoming too varied, complex, and significant for one organization to do it all well. Similarly, the notion that, in the age of information, all communications flows in societies operating largely on the market principle would pass through one streamlined super-pipe of a single organization is hard to entertain on technical, economic, or political grounds, except by reference to the present balance of political power. But these conditions are not very likely to prevail, as has been argued above. Once the notions of the centralized network are breached in some respects, the process is hard to contain.

The process is inevitable, not because it leads necessarily to a superior result, but because the centralized network is an anomaly, although one too familiar to notice. As long as the economic system of Western industrialized democracies is based on markets and private firms, the exclusion of major economic parties from a major field is an unstable affair. It is hard to maintain a dichotomy between telecommunications and the rest of the economy. The reference to infrastructure service is too vague to be useful here. Telecommunications, unlike a lighthouse or a road, is not a public good in the classic sense; users can be excluded and charges can be assessed, breaking the two conditions for public goods. Nor do the externalities of network participation require a hierarchical system; subsidies can be at least as effective.

Thus, the telecommunications field in the future will more closely resemble the rest of the economic system and will be less a part of politics. It may be much more complex, and perhaps less efficient in some ways than the old system, but it will be a truer reflection of an underlying complex society.

» References

Codding, George A., Jr., and Anthony M. Rutkowski. 1982. *The International Telecommunication Union in a Changing World.* Dedham, Mass.: Artech House.

Dallmeir, Martin. 1977. *Ouellen zur Geschichte des Europäischen Postwesens 1501–1806.* Kallmuenz: Verlag Michael Lassleben.

Holcombe, A. N. 1911. *Public Ownership of Telephones on the Continent of Europe.* Boston: Houghton Mifflin.

Noam, Eli M. "The Political Economy of ISDN." Working paper, Columbia University, Center for Telecommunications and Information Studies.

Stephan, Heinrich. 1859. *Geschichte der Preussichen Post.* Berlin: Verlag der Königlichen Geheimen Ober-Hofbuchdruckerei.

2

» The Globalization of Telephone Pricing and Services

Peter Cowhey

In the last fifteen years, a global "greenbelt"—a loop around the world of financial centers tied together by the dollar—has created a 24-hour financial market, along with supporting services. The emergence of this greenbelt has fundamentally changed the political and economic environment for telecommunications. It is no accident that the three countries with the most advanced experiments in telecommunications competition (the United States, the United Kingdom, and Japan) are also the world's preeminent financial centers. This troika has had to rationalize its information and communications systems to reduce costs and create new services.

This phenomenon can be understood in another context. Until recently, services constituted over 50 percent of the gross national product (GNP) in industrial nations, but they were organized on a national basis and were relatively closed to foreign competition. Meanwhile, manufacturing became thoroughly multinationalized, and these manufacturing firms demanded global services. The financial sector led the service sector in responding and underwent its own internationalization. Then other service industries in the money centers (such as law) had to respond. Today, there is a ripple effect as the reorganization of all services in large financial centers forces the service firms throughout each country to restructure and internationalize. This will eventually happen to local exchange carriers (LECs), even though they operate primarily at the local or state level. The LECs have to begin planning for the transition today.

» *Reform Experiments*

In a larger sense, the global story of regulatory experimentation with competition in telecommunications is also a tale of political and economic struggle.[1] Many of the traditional cross-subsidies attached to the telephone industry are under attack or are disappearing. Several strategic factors are common to the struggle over regulation everywhere. In all countries, the telephone system supported a wide variety of cross-subsidies: the support of national equipment manufacturers, local residential services, the postal system, and the publishing and media industries. LEC pricing reflected these commitments.

The fundamental reason for regulatory reform was that over time information and communication become more vital to a relatively small set of players in the world market. At the risk of oversimplification, three sets of players can be singled out—larger users of long-distance services, computer manufacturers, and the publishing and media industries.

A relatively small number of users have a large stake in the organization of communications systems. Approximately 5 percent of domestic and international long-distance users constitute over 50 percent of the long-distance traffic. Because communication and information services have become more salient to their production and profit strategies, the long-distance users are an easily organized and influential class of players in the regulatory system.

A second phenomenon is the intersection of the telecommunications industry with computers. This forces the entry of the computer industry into the telecommunications regulatory arena on a large scale. The semiconductor and computer revolution brought a whole new generation of electronics companies to the fore as dynamic competitors. They could not afford a continuation of the strategic support that the old monopoly system gave to traditional telecommunications suppliers such as NEC, Fujitsu, or Western Electric.

These new electronics firms in the United States and Japan will lobby in all countries for competition and for restructuring the architecture of enhanced communications services.

Finally, the publishing and broadcasting industries recognized that the expansion of their markets required movement into specialized electronic media (such as videotext and "narrowcasting" specialized programs). They believed that it would require a revolution in the pricing and delivery of communications services to make this possible.

» *Effects of Internationalization*

Many believe that telecommunications deregulation in the United States constituted unilateral disarmament. The United States opened its equipment market while other countries kept theirs closed. This country changed from a net exporter to a major importer of communications equipment and is now hurriedly experimenting with different ways of opening foreign markets. All of this is true, but it does not reveal a more fundamental development—that the equipment industry itself is being profoundly internationalized.

It will cost $2 billion to develop the next generation of central office switching systems. Not a single national market in the world (including the United States) can support this developmental cost. As the global telephone equipment market goes through a major shakeout and reorganization, the companies with strong international access will survive. For example, Siemens cannot be a major switch provider in the world unless the company has at least 5 percent of the U.S. market and some percentage of the Japanese market. The pressures will increase for overseas equipment manufacturers to become involved in local exchange regulatory politics because these politics influence the purchase of switches. So, the internationalization of the equipment market leads to the internationalization of the political pressures on LECs. And the choice of switch influences strongly the software system that defines many LEC service options. For example, international carriers may prefer to work with local carriers that use the same type of central office switch because similar switches make it easier to harmonize services.

Telecommunications services within the United States are also becoming internationalized. The beginning of this process is exemplified by NYNEX's effort to invest in PTAT, the trans-Atlantic fiber-optic company that will provide private circuits into Europe.[2] NYNEX also is negotiating with the Dutch communications authorities to create an international value-added service between the United States and the Netherlands within the boundaries of the MFJ (modified final judgment) consent decree. Both moves show how a regional Bell operating company (RBOC) can influence traffic loads on its domestic network by paying closer attention to its international environment.

Foreign companies also are preparing their entrances into the United States. In the recent U.S.-Canada free-trade negotiations, Bell Canada persuaded the Canadian government to propose (unsuccessfully) that it be allowed to enter the voice resale and perhaps the communications facilities businesses inside the United States. (Bell Canada had already bought a natural gas pipeline company in the United States, a purchase that gave it valuable rights-of-way.) This is only one small example of the potential influence on domestic traffic of emerging global voice networks. Both British and French firms have stakes in the U.S. long-distance market. The U.S. government also will eventually sign agreements with Japan and the United Kingdom that will authorize reciprocal provision of value-added services by each country's companies on a fully competitive international basis. The entry of British, Japanese, and other foreign firms into certain segments of the U.S. domestic market is almost assured. The Japanese may well have a competitive advantage on some value-added services because the costs of their electronic components are 15–20 percent lower than the costs incurred by their U.S. counterparts.

» *Implications for LECs*

Virtually every major country is experimenting with ways to reconcile new approaches to regulating domestic communications and the reform of international rules governing the provision of international communication and information services. Even some of the countries most committed to retaining domestic communications monopolies are finally creating independent regulatory commissions to limit the traditional authority of the established telecommunications administrations. They also exhibit renewed interest in reforming prices by adopting Ramsey-style pricing of services in order to avoid bypass.

There are also a variety of experiments in structuring for greater competition. If the U.S. formula for competition can be simply stated as, open markets except for local basic services, then the basic formula in Japan and the United Kingdom can be described simply as, competition in all services (local and long distance, basic and enhanced), but with a limited number of entrants into the basic facilities markets. By controlling the number of entrants into the basic facilities markets, the government can indirectly push the firms to keep a control on the amount of price-cutting and, therefore, indirectly keep cross-subsidies for local customers (including the accellerated provision of new services) out of the companies' cash flow. The result is that, for example, price cuts by new Japanese carriers have been limited to approximately 25 percent.

Another method for increasing competition has been proposed for adoption by the governments of Canada and West Germany. This formula mandates that the basic facilities for telecommunications remain a monopoly in order to keep economies of scale in the physical plant, but that all services may eventually be open to competition (although both stick initially to enhanced services).[3]

The final method is the limited value-added approach. This path recalls early U.S. efforts to restrict competition largely to enhanced and selected private-line services. But having watched the U.S. frustrations with what constitutes a basic or enhanced service, other countries operate on something closer to the principle that what is not explicitly permitted is presumably forbidden (instead of attempting to create a blanket definition for enhanced services). These openings have prompted IBM to create value-added networks in France and Italy under government approval. RBOCs are becoming partners in cellular phone networks in such countries as France and Argentina.

Each of the regulatory experiments throughout the world carries different implications for pricing, the definition of services, competitive options, and the organization of international telecommunications. For example, the world might move toward more competition in communications but adopt ground rules that are more hospitable to Japanese than to U.S. competitors. To discourage such a development, the U.S. government has recently made a determined effort to influence the process of reform. Two fundamental sets of international negotiations are occurring simultaneously.

The first is inside the General Agreement on Tariffs and Trade (GATT). The advocates of regulatory reform want to transform telecommunications and other services from regulated industries (which usually are treated in specialized intergovernmental forums) into trade commodities. To over-simplify, this effort could eventually turn making international telephone calls into the equivalent of shipping television sets across national borders. Then, in theory, the services would be under the auspices of the world's trade organization, GATT, a body that is more sympathetic to competition than the ITU (International Telecommunication Union) and is dominated by trade ministries with far broader constituencies than traditional communications ministries. Such a change would be very much in the interests of U.S. service firms. The problem is that GATT has no provisions for regulating trade in services. The new round of GATT negotiations will have to create this authority, and so far it is easiest to focus on enhanced services and the rights of users of the communications system. Then something will have to be done to improve the expertise of GATT participants in the intricacies of communications. Inevitably, this will require some sharing of work with the ITU.

At the same time, the World Administrative Telegraph and Telephone conference in 1988 updated ITU regulations to make them apply to the brave new world of computer and information services. The United States firmly opposed the draft regulations that threatened to extend international regulations (and perhaps common carrier obligations) to newer service offerings. According to U.S. views, the ITU is a good place to handle technical work on communications policy, but its commercial regulations are anticompetitive. While the United States accepted in principle limited competition in voice services, it argues that ITU regulations should not cover enhanced services and their providers. It also calls for recognition of the right of countries to set up much more competitive arrangements for all services and facilities by means of special bilateral agreements.

No U.S. official knows where all this activity is going, although it is obvious that action is imperative. However, the effect of the negotiations will eventually be to internationalize domestic regulations by implementing the principle of "market access" by means of new international trade rules governing both international and domestic telecommunications systems.[4] Countries will claim the right to experiment freely with regulatory reform but will acknowledge that domestic regulations must accommodate the minimum requirements stipulated in the international trade agreements. While international rules accommodate substantial latitude in interpretation, this international arrangement is a fundamentally new challenge for domestic policymakers.

What form will these new obligations take? One feature will be the creation of rights for both global suppliers and global users of communications services. (In many cases, companies are both buyers and sellers.) In local communications regulations, the user community is already represented and has rights, but this is almost unheard of in the international area. When foreign companies operate in the United States, they will have a minimum set of internationally defined rights. If local pricing and regulations are not in harmony with the international agreements and their contents, the local regulations are subject to challenge. So, when foreign users establish a presence in the local exchange service area, they will come with a vested bundle of rights that will not be defined by the local regulatory commission. (It is highly likely, for example, that U.S. LECs will have to provide interconnection with technical protocols common elsewhere in the world but not central to the U.S. networks.)

There is always substantial uncertainty about what complex international negotiations will yield. If the United States succeeds only in its most modest proposals, global negotiations will concentrate on the rights of users and suppliers regarding two specific market segments—enhanced and private-line services. Exclusion of switched voice services from trade talks would produce a very biased vision of the global regulatory system, which will be transported down to the level of the local exchange.

Another fundamental problem in these negotiations is how to monitor good faith and compliance of all countries with these new rules. Monitoring in an international situation is very difficult and leads to a preference for very robust but simple-minded rules. For example, the U.S. free-trade agreement with Canada contains provisions that build a strong presumptive right to flat-rate, full-period leased circuits for users. Many LECs advocate a movement toward usage-sensitive services. The ability to do this with regard to international customers in the local exchange, however, is being constrained by international agreements.

Trade talks will certainly have other implications for the organization of LEC communications systems. For example, the rules will influence how international traffic feeds into the local loop. Private international satellites will make many old "gateways" to the local loop less attractive. The negotiations also will have implications for integrated services digital networks (ISDNs) in terms of what an LEC can or cannot do. However, it is still not clear how the interests of LECs will be represented in the formulation of a U.S.

negotiating position. The RBOCs sit on industry advisory committees for U.S. trade negotiators, but coordination of international and
domestic planners is weak within the RBOCs. Just as fundamentally, internationalization of the local loop will deepen tensions between larger and smaller LECs, and the latter have been silent on
trade talks. The smaller LECs may have to court foreign carriers
whose U.S. resale networks could be vital to them. All the factors
that have been discussed could change the LEC business significantly
over the next five to seven years, and it is not clear that the LECs
have begun to think and act on them sufficiently.   .

» *References*

Aronson, Jonathan David, and Peter F. Cowhey. 1988. *When Countries
    Talk—International Trade in Telecommunications Services.* Cam-
    bridge, Mass.: Ballinger Publishing Company.

» *Notes*

1. This paper draws on data and analysis presented in greater depth
   in Aronson and Cowhey (1988).
2. NYNEX has been frustrated in its quest for regulatory approval
   of its participation in PTAT. Its investment option has lapsed,
   although its pursuit of permission has not. Pacific Telesis has obtained an equity share in the Japanese end of a private fiber-optic
   cable connecting the United States and Japan.
3. Neither Canada nor West Germany will permit competition in basic
   services. Many observers believe that Canada will eventually permit competition in long-distance voice services and that, in the next
   decade, West Germany will allow competition in voice packages
   that are part of a broader mix of data and video.
4. These claims rest on examination of interagency working papers
   circulating in the U.S. government during the summer of 1988.

3

» *Regulatory Practices and Access to Capital Markets*

Larry F. Darby

» *Introduction and Summary*

Regulatory practices in the telephone business are related in a variety of ways to the cost of capital and other aspects of carrier access to capital markets. Channels of cause and effect run in both directions between regulatory arenas and markets for carrier securities. These remarks will focus on some key, but frequently neglected, elements of this mutual interaction. In particular, the discussion will focus on the ways in which regulatory change can and does influence the cost of capital to a regulated telephone company.

This paper will begin with a discussion of traditional investor views of the telephone business, before discussing the major events whose combined effects have been to alter dramatically the traditional view. These major disturbances—divestiture, diversification, and deregulation—have twisted the risk/return profile of telephone securities, changed investor expectations, and thereby altered the outlook for carriers' cost of capital in the long run.

In this new financial, economic, and legal environment, the effect of different regulatory practices influencing carrier rates has been and will become more instrumental in determining the long-run cost of capital.

From investors' points of view, the details of common carrier regulation matter, and they matter a lot.

» *Traditional Views of Telephone Securities*

Investors in U.S. telephone securities are motivated by the full range and diversity of human and institutional incentives. However, a common motivation of investors is a desire to enjoy growth in wealth while being assured of high levels of safety and protection of initial outlays. Thus, investors seek large returns and low risk. These two incentives are in conflict, since the range of available investment opportunities generally is characterized by higher returns being paired with higher risk.

Investors must balance and trade off these two competing considerations, in accordance with the dictates of individual investor preferences.

Investors may buy either telephone debt issues or telephone stocks (equities). In either case, they expect to see their original outlay grow in value through periodic payments in the form of interest or dividends and through appreciation in the market value of the security (capital gains). Investors look for security, predictability, yield, and growth.

Historically, telephone securities have enjoyed a very special place in investors' portfolios, owing to the almost unique characteristics investors have perceived in the quality of telephone paper.

While not nearly so "risk-free" as government securities, telephone issues have been regarded as carrying very low financial risk, market risk, and operating risk. There is virtually no threat of bankruptcy and—to this author's knowledge—there have been no cases in which a carrier was unable to meet its fixed, long-term obligations. Companies have historically maintained very conservative capital structures—less than 50 percent debt and no preferred stock. Earnings have almost invariably covered fixed charges by comfortable multiples, while construction budgets have been largely funded by internally generated cash flow from depreciation charges and deferred taxes.

Of course, this is a very capital-intensive business, and the consequently high ratio of fixed to variable costs suggests the potential for substantial operating risk. However, this potential has been substantially offset by the historically negligible market risk, occasioned by the virtual absence of market competition for most of the revenue.

Income streams from major lines of business, as well as in the aggregate, have tended to be both stable and reasonably predictable. Revenues have been insulated from the vagaries of the general business cycle. Market demand has a fairly low income elasticity; receivables are of extraordinarily high quality; and dividends have been regular, reliable, and steadily growing. The pace of the introduction of new technology has been both measured and predictable. There have been no great surprises or substantial changes in costs over short periods of time.

The industry has been widely regarded by investors as enjoying "reasonable," if not laudable, treatment by its major government overseers—various state public utilities commissions (PUCs) and the Federal Communications Commission (FCC). Most major regulatory changes have been more or less predictable, and their effects on market valuations have generally been discounted in significant measure well in advance of actual regulatory decisions. There have been no abrupt and major share valuation "shocks" brought about by regulatory fiat. This is in sharp contrast to some regulatory decisions in the transportation and energy industries. Prior to divestiture, there were no major regulatory or judicial surprises. Investors, with some notable exceptions, have generally given telephone regulators deservedly high marks.

Investors' valuations of telephone stocks have been very sensitive—and not surprisingly so—to the movement of long-term interest rates. The historic security of telephone companies' earnings has permitted their stocks to trade much like long-term bonds, with the same kind of sensitivity to the movement of interest rates. When interest rates are low or falling, bond prices are high or rising, thereby exposing investors to capital loss. In this environment, utility stocks look relatively attractive, and telephone stocks particularly so. By similar reasoning, telephone stocks have tended to suffer in less favorable interest rate environments.

Most analyses have shown high and significant relations between interest rates and telephone stock prices over time. Investors have apparently regarded telephone stocks and long-term, high-quality bonds as reasonably close substitutes, with the terms of trade between the two a function of market rates of interest.

In short, the traditional view of telephone stocks has placed them in that select group of "blue-chippers" recommended unabashedly for the portfolios of widows and orphans whose requirements include moderate, but very secure, earnings and growth.

The traditional investor view of telephone securities has been very supportive of one of the major goals of common carrier regulation. Telephone companies, with very few exceptions, have had excellent access to capital markets, as indicated by their generally superior credit ratings, stock prices reflecting high earnings multiples, and the consequent, relatively low average cost of embedded capital. Capital costs are, of course, an important part of a firm's revenue requirement. Depending on the company and the methodology, capital costs constitute around 20 percent of the total revenue requirement in a given year.

» Shocks to the Traditional View

Several developments have combined in recent years to modify the basis for traditional investor valuations of telephone securities. It is helpful to group these changes under the headings of deregulation, divestiture, and diversification. None of these is a single event. Rather, each is an evolutionary process that is steadily eroding the basis for the traditional view of telephone securities.

Taken together, the forces embodied in these processes have dramatically altered the financial and commercial environment within which telephone companies operate and, on balance, have significantly reduced the overall comfort level of investors in telephone and related securities.

Investment patterns, operating and pricing practices, corporate strategies, financial and corporate structures, lines of business, and, indeed, the most fundamental rules of the regulatory game look very little today like they did less than a decade ago.

Deregulation

The rhetoric of deregulation has far and away dominated its practice —at least insofar as it applies to incumbent, so-called dominant carriers. Nevertheless, changes in the application of traditional regulation over the past decade have been sufficient to force investors to review long-held notions about telephone securities.

It is helpful to think of deregulation as encompassing several important categories of change. These are:

1. Permitting newcomers to enter previously foreclosed common carrier markets
2. Tailoring regulation or forebearing regulation of new entrants
3. Permitting incumbents to enter new markets
4. Changing or designing rules for incumbent competitive behavior in traditional and new markets

Complete deregulation, in the sense that the full panoply of common carrier regulation has been discontinued (for both newcomers and incumbents) in a particular line of business, has occurred most conspicuously in the terminal equipment business—manufacturing, installation, maintenance, and repair.

In other submarkets, partial deregulation has generally opened market entry by lifting regulatory barricades and has been characterized by regulatory forebearance with respect to new entrants and by limited modification of several of the broad spectrum of regulatory rules governing incumbent carriers.

The more powerful of these incumbent carriers, the dominant carriers, have enjoyed very limited deregulation with respect to either their ability to respond in their own markets to new competitors or their discretion to enter new markets. The regulatory structure governing the conduct of incumbent carriers is not markedly different from what it was a quarter century ago. The changes that have been made are more in the nature of fine-tuning than of a fundamental overhaul.

Open entry and the asymmetric incidence of deregulation have dramatically increased market risk to incumbents. Of course, it could not be otherwise in markets that had previously been monopolies virtually guaranteed by federal and state governments. Easing or eliminating regulatory barriers to entry has created, destroyed, and redistributed risk and financial value among various market contenders.

The list of newly competitive lines of business is rather long, stretching from terminal equipment through toll and a variety of enhanced or value-added services to yellow pages and operator services. The weighted average risk of assets deployed and the cost of capital used in these markets are significantly higher than otherwise would have been the case.

In short, open entry and increased competition have increased the average cost of capital deployed in the industry. New competitors of the old Bell System frequently have higher capital costs than the incumbents, although there are some exceptions, and the cost of capital to incumbents is higher by amounts reflecting the added market risk.

Divestiture

The impact of divestiture on the cost of capital to the telephone industry is somewhat less clear-cut than the impact of deregulation, as defined above. Different aspects of divestiture cut different ways. Some features of divestiture tend to increase the cost of capital, while others tend to reduce it. On balance, however, it is probably the case that divestiture has increased the overall cost of capital to the former partners in the old Bell System.

From the perspective of risk-averse investors, the single most important effect of divestiture has been the reduction of market power of the old Bell System. Taken together, the piece parts after divestiture are significantly less dominant in the marketplace than they were before. A very important reason for this is the fact that divestiture created the opportunity for market competition among the former Bell siblings.

Prior to divestiture, there was very little, if any, market rivalry among the various earnings centers. Beyond that, from the point of view of an AT&T shareholder, any market rivalry was approximately a zero-sum game. Thus, it did not really matter to an AT&T shareholder how jointly produced toll revenues were divided between Long Lines and one of the wholly owned operating companies. Nor was there a lot of investor concern about, or even attention to, the apportionment of common costs among various operating companies under the various license contract arrangements.

Each of these processes was merely shifting costs of revenues around among various pockets in the same pair of corporate trousers.

In the postdivestiture market environment, the rivalries among the former partners have increased enormously. Most importantly, the rivalries and the distribution of network costs and revenues are no longer internalized under a single set of stockholders.

Despite the line-of-business restrictions on the former partners, there is a large and increasing domain of actual and potential competition among them. For example, AT&T competes with the operating companies in the sale and service of terminal equipment, in some intraLATA (local access and transport area) toll markets, and for shared tenant services—and has the ability to bypass, and thereby to compete with, the local exchange network in serving large users. Competition among the operating companies themselves is becoming increasingly widespread and intense. They compete, for example, in the terminal equipment business, in the extremely lucrative directories business, and in the cellular telephone business. Thus, divestiture and developments since then have combined to increase market competition and market risk, thereby creating significant upward pressure on the cost of capital relative to the status quo before divestiture.

As important as divestiture has been in creating new domains for competition, it has also rendered even more unstable the historic channels of "cross-subsidy." Open entry and deregulation of new entrants effectively undermined the political and financial basis for equilibrium in subsidy flows among carriers, users, services, and parts of the country. Divestiture will in the long run combine with the FCC's procompetitive policies to destroy completely these long-standing, social ratemaking practices.

In the short run, however, investors perceive greater uncertainty as regulators argue about the necessity of and means for converting rates to a cost basis. The uncertain, and some say glacial, pace of the transition toward cost-based rates leaves some very large revenue streams subject to "uneconomic" diversion. And the regulatory response at both the state and federal levels is simply not clear to investors.

The role of divestiture in all this has been to destroy the basis that made large transfers of revenues politically and financially viable. Jurisdictional separations, toll settlements, and division of revenues all depended in some measure on AT&T's common ownership of the dominant share of the long-distance and local networks. Notwithstanding the rhetoric about economic efficiency, cost causation, and all that, the reason for the system of residential end-user access charges is divestiture—pure and simple. The separation of Long Lines from the operating companies converted an intracorporate transfer of revenue into an intercorporate transfer. For AT&T shareholders, the result was to externalize an internal cost.

Similarly, the Unity 1-A Agreement, which is an agreement among virtually all the LECs that addresses geographic subsidies, is largely a product of divestiture, even though there have always been some disequilibrating forces at work. Cost-averaging and revenue-pooling according to long-standing AT&T and industry practice have resulted in substantial transfers of revenue from one section of the country to another. Divestiture, per se, did not change those practices. Divestiture did, however, make the process more visible, and more importantly, divestiture changed what had previously been a stockholder "wash" into a situation where some stockholders gained at the expense of others. Again, internal costs and benefits of an integrated Bell System were externalized by divestiture.

Forces for rerouting or drying up several other subsidy streams were unleashed by divestiture. Over the long haul, it is very unlikely that yellow pages, for example, will be able to continue to generate current levels of contribution to the local company revenue requirement. The business has generally gotten more competitive since divestiture, and absent any significant barriers to entry and product differentiation, it seems quite likely that competition in the directories business will become even more pervasive and intense in the future.

The list could go on. But the point here is that divestiture has reinforced the trend toward cost-based pricing launched originally by FCC decisions to open entry into selected markets. The transition to cost-based prices provokes uncertainty among investors, even though they may feel more secure once the target of cost-based prices has been achieved. In the meantime, however, there is considerable investor uncertainty about the security of large revenue streams and the willingness of regulators to permit carriers to protect them.

Before turning to the effects of diversification on the cost of capital, a brief mention of what some have called the "reverse portfolio" effect of the divestiture on the cost of capital to the former Bell System companies is in order. The old Bell System was a formidable force in the capital market. In addition to its monopoly power in markets for its products and services, the Bell System exercised some monopsony power in its demand for capital (credit in particular). Its sheer size and the frequency and regularity with which it came to the capital markets undoubtedly gave it some favorable terms of access, which were sacrificed at divestiture. Beyond its size, however, AT&T enjoyed enormous diversification with respect to its geographic and product lines of business. If technological developments favored telecommunications products at the expense of telecommunications services, or vice versa, AT&T's diversity provided a hedge. Similarly, AT&T's geographic diversification in some measure insulated its securities from the vagaries of world energy markets inasmuch as regional effects in the industrial, energy-using Northeast were offset by impacts in energy surplus areas in the Southwest.

The benefits of this portfolio effect were manifest in lowered risk from the investors' point of view and were reflected in AT&T's cost of capital. These benefits were, for the most part, given up at divestiture. Though investment funds have been initiated that try to reconstruct the old Bell System securities, they have not been successful in reconstructing its very favorable risk profile. And, of course, none of the former partners can obtain debt financing on the combined strength of the group.

To sum up, divestiture has had the effect of increasing the cost of capital to the members of the old Bell System. Although there is almost no way to make reasonable estimates of the amount of added cost, there is little doubt, that higher capital costs offset in some measure whatever benefits have been derived thus far from divestiture. The good news is that the additional capital costs attributable to divestiture should diminish over time as the new entities mature and time solves many of the associated uncertainties.

Diversification

There has been much discussion in regulatory circles about the effects of telephone company diversification on the cost of capital. In principle, of course, almost any change in the composition of a company's assets and output will alter the risk/return profile of the company's securities, thereby influencing the company's cost of capital. In practice though, it is difficult to document the effects of diversification on the cost of capital. In the case of the former Bell partners since divestiture and in the context of deregulation, it is not clear just how much diversification has mattered one way or the other.

When one looks at specific cases, it is somewhat surprising to discover that the Bell companies are not really very diversified. Any reasonable index of diversification should reflect the degree of difference between the core and new lines of business, as well as the revenue (or earnings) contributions of the new lines in relation to the contributions of the core business. As it turns out, the companies are not notably diversified in either of these dimensions. There are several reasons for this.

In the first place, there are quite severe legal restrictions on the discretion of the companies to diversify, embodied in the modified final judgment (MFJ). History will judge the wisdom of those restrictions as administered by the decree court, but there can be no doubt that they have restricted the range of BOC (Bell Operating Company) diversified activity. The line-of-business restrictions do not look all that complicated. The application of the restrictions on manufacturing, interexchange services, and information services has, however, uncovered a wide range of enterprise that is simply foreclosed to the BOCs. Numerous lines of business in which the BOCs could make a contribution contain some small and frequently uncertain elements of the proscribed activities, which effectively "poison" the entire line of business or a particular acquisition.

If a particular diversifying initiative does pass muster with the decree court, the proposal must still clear regulatory hurdles at both the federal and state levels. If the initiative raises questions of potential cross-subsidy or possible exploitation of the "bottleneck" (and almost all of them do), there are still further limitations likely to be placed on diversification.

Much of the business activity that is regarded by regulators as diversified activity is not all that different from what the companies have done historically. The difference is that the business is not regulated; or, it is regulated differently from the core common carrier business. Businesses that fit this category include nonmanufacturing aspects of the terminal equipment business, the directories business, cellular communications, consulting, real estate, computer/software services, and the like.

It is not by accident that the companies have not diversified far from their core businesses. Each of the companies has emphasized its intention to limit diversification to lines of business that it understands and that add to or derive value from the core business. Thus, none of the companies is remotely approaching the diversification patterns of the major conglomerates—ITT, LTV, and others—that materialized during the sixties and seventies. Indeed, the BOCs are not as diversified as some of the larger independent telephone companies.

Revenues from diversified, nonregulated lines of business are still quite modest relative to the incomes generated from the core businesses. This is due in some considerable measure to the simple arithmetic of compounded growth. The managements of several BOCs have announced intentions to grow diversified or unregulated revenues to 25 percent (or 50 percent) of total revenues by some future date certain. Starting with a very small base of unregulated revenues requires enormous annual growth rates just to keep pace with a base of regulated revenues that is an order of magnitude larger, but growing at modest rates. To illustrate, the annual increment to the core business of a "typical" BOC creates a new revenue stream roughly the size of Comsat's. Thus, the BOCs are primarily local exchange companies and will be for some time, barring some enormous acquisitions and/or relaxation of the line-of-business restrictions. That character will change, but only slowly over time.

Although the companies are not highly diversified at present, they are clearly more diversified than at divestiture and are destined to be even more diversified in the future. The threshold question is the impact of diversification on the cost of capital. The regulated core business dominates the unregulated, diversified parts so completely that the overall risk/return profile of the BOCs has not yet been significantly transformed by diversification. Though the impact of BOC diversification on the overall cost of capital to date has probably been negligible, that situation will most assuredly change in the not so distant future as the companies' diversification plans mature. Nevertheless, for the time being, the cost of equity to the holding company may well be a reasonable proxy for the cost of equity to an operating company (for ratemaking purposes).

» *Regulation and Access to Capital*

Having touched briefly on the capital market effects of the broad trends of deregulation, divestiture, and diversification, the remainder of the paper will discuss the potential impacts of some illustrative regulatory issues.

There are several issues on today's regulatory agendas that will influence securities analysts' (and hence investors') views of telephone securities. Bear in mind that there is a very diverse set of telephone securities from which investors may choose. While most of the preceding comments have been focused on the former Bell System companies, investors have a keen interest in a wide range of corporate assets that either compete with or complement the output of the Bell companies. Thus, a regulatory negative for a Bell company may very well be a significant positive for one of its competitors or potential competitors.

Some may recall a company—or would-be company—called

DATRAN. DATRAN never transmitted a bit of data. During the gestation period of DATRAN, there was considerable uncertainty about the extent to which the FCC would permit AT&T to respond to new competitive entry by lowering its data rates. The verdict in the capital markets was that DATRAN was not likely to survive the rate responses permitted AT&T by the FCC. That prophesy became self-fulfilling.

The DATRAN story illustrates vividly the influence of regulatory decisions involving ratemaking on the flow of capital into the industry. That influence continues today. Potential new entrants prowl financial districts in pursuit of venture capital to fund entry into "niche" or other markets. A common question confronting all would-be competitors is the likely response of incumbents.

The pricing flexibility accorded incumbents is important in several respects. The FCC has several dockets in which the rate response of incumbents to new entrants is at issue. Each of the state regulatory bodies probably has a couple as well. These are important dockets to investors. Their resolution will influence the market risk and potential loss of market share, income, and earnings for incumbents. Simultaneously, of course, these dockets will determine the growth, security, and, indeed, the survivability of competitive firms.

In this respect, regulators are regarded by some in the financial community as market allocation managers whose decisions dramatically influence the outcome of market processes. Securities analysts follow regulatory proceedings very carefully for clues to the future security, growth and earnings potential of telecommunications assets. Regulatory uncertainty, that is, uncertainty about future regulatory rules of the game, creates regulatory risk for all contenders in the marketplace. This uncertainty may translate to increased cost of capital all around.

Securities analysts are both amused and befuddled by the intense, historic attention of regulators to cost allocation rules. They know well that the allocation of common costs is quite subjective, but nevertheless quite powerful in its ability to determine the growth of new entrants and incumbents alike. They understand the regulatory signals embodied in accounting cost allocation rules. Fully distributed costs mean that new competitors will grow and incumbents will lose market share. They realize that long-run incremental cost will limit the growth of competitors. And the cost of capital to various contenders may be significantly influenced by investors' perceptions of the future, which they draw from cost allocation rules.

Regulatory rules regarding the evolution of rate structures are very important to investors. The reason is simple. The structure of rates influences the security of various revenue streams, revenue growth opportunities in various service lines, and more generally, a firm's overall competitiveness or prospects for survival. For example, usage-sensitive rates promise greater opportunity for revenue growth than do flat rates. They lessen competitive exposure to loss of markets to resellers. But they also promise greater variability (therefore risk) in various income streams. On balance, usage-sensitive rates are regarded by investors as a positive development.

In addition to pricing flexibility and rate structure questions, the cost of capital is sensitive to other regulatory issues as well.

Capital recovery has in recent years become a concern of investors. The concern derives in large part from sentiments expressed by some regulators who would respond to any deficiency in depreciation reserves by simply requiring investors to write down the value of their assets. Investors will not be as willing to finance major renovations of the public switched network through fiber optics and high-capacity digital switches if regulators require them to write off past investments in wire and earlier generations of switches.

A fairly recent regulatory issue has emerged that could have significant consequences for the future cost of capital to regulated carriers. The issue has arisen in several contexts but can be phrased as a simple question: Who owns carrier profits? A variety of state regulators and Judge Harold Greene, among others, have raised the issue in the context of various business proposals and plans by the RHCs. The term "ratepayer funds" keeps cropping up as a description of carrier earnings or cash flow from other sources.

Now stockholders believe that they—not ratepayers—own the earnings from regulated telephone assets. Investors recognize and respect the discretion of regulators to limit the use of those earnings in ways consistent with regulatory perceptions of the public interest. However, it is important to remember that any encumbrances on the reasonable use of telephone company earnings, or any extraordinary claim of regulatory control over their future disposition, dilutes the value of those earnings to investors. In a competitive capital market, that can only put upward pressure on the cost of capital to the companies involved.

» *Conclusion*

The influence and role of capital costs in ratemaking and other regulatory processes is well documented and the subject of detailed analysis and broad discussion. The purpose here has been to take a preliminary look at channels of causation running from regulatory arenas to capital markets and in particular, to sketch out some of the sensitivities of capital costs to the resolution of specific kinds of regulatory issues.

Much more needs to be done to explore these relationships. There is little in the literature analyzing the effect of different forms of ratemaking rules or regulation more generally on capital costs.

It is quite clear that regulatory structures, processes, and decisions have the potential for substantially influencing the long-term cost of capital to the industry. That is significant. Of much greater significance in such a capital-intensive industry, however, is the fact that the industry in general, and its regulators in particular, know so little about those impacts.

» *Postscript on Price Caps*

The core of this paper was prepared in 1987. While the basic analyses have not changed since then, some key regulatory events have intervened.

During the second quarter of 1988, the FCC formally proposed a radical change in its method of controlling the overall rate level of regulated and dominant common carriers. Specifically, it proposed the replacement of rate base/rate-of-return regulation with a regulatory scheme that would permit rates on average to change in accordance with an index reflecting both changes in carrier input prices and productivity. Thus, under this FCC proposal, changes in the level of rates would no longer be constrained by changes in the weighted cost of equity and debt capital.

The FCC price cap proposal would also introduce a dramatically new basis for regulating the rate structure. At present, dominant carrier rates are, for the most part, based on accounting costs allocated according to a variety of regulatory conventions. This new proposal would allow rates for services within a given class, or "basket," to vary at the discretion of the carrier within certain "bands" or ranges specified by FCC rules.

The FCC proposal has provoked quite a bit of controversy, and the timing and extent of its implementation are not at all clear. However, several aspects of the proposal have made it potentially attractive to an investment community that has become increasingly wary of government utility regulation and promises of regulatory change.

Some opponents of the FCC proposal have used the fact of potential investor support as an argument against its adoption on grounds that if investors favor it, then it must be bad for consumers. It is, of course, naive to regard the price cap proposal as a zero-sum game between investors and users. But the argument illustrates, nevertheless, some widely held notions about the impact of regulatory change and investor reactions. If the proposal leads to greater economic efficiency, as most analysts expect, then its implementation will generate positive surpluses that the regulators can distribute among producers and consumers, thereby making both groups better off.

Earnings will not necessarily be higher under the commission's proposal than they would be under rate base/rate-of-return regulation. Investors have reacted positively to the prospect that earnings *might* be higher, even though there is continued risk that they may well be lower than the cost of capital. The latter risk is present under the current scheme of rate-level regulation inasmuch as carriers are not guaranteed that they will earn the cost of capital, but only permitted to do so.

Much of the reaction of investors to the price cap proposal is a reflection of investor aversion to regulatory risk manifest in the biennial roll of the dice known as rate-of-return proceedings. These proceedings are fraught with uncertainty, delay, and no small amounts of arbitrariness and ambiguity. Despite the trappings of scientific inquiry and impartial analysis, rate-of-return proceedings are political to the core, and investors know it.

Some investors regard price caps as a regulatory development that will permit them to evaluate and compare different carrier managements and to make them more accountable to the owners of the assets. Under the present regulatory scheme, owners do not reap much advantage from increases in managerial, operating, or other forms of efficiency. They get paid pretty much the same irrespective of the skill of management in controlling or reducing cost. If regulation works in practice according to the theoretical principles underlying it, then the surpluses created by increased efficiency over time will be captured by regulators and converted to lower rates for users. Investors understand the asymmetric payoff structure under rate base/rate-of-return regulation. And the terms they attach to the supply of capital to the industry reflect that understanding.

A final source of favorable investor reaction to the FCC proposal may well be the recognition that the constraints on carrier pricing of metaphysical exercises of regulatory cost accounting will be lifted, and the rate structure will be freed to seek its own correspondence with the structure of economic costs. The pricing flexibility accorded carriers in the FCC plan will alleviate investor concerns that regulatory commissions have come increasingly to view themselves as market managers. Investors understand how market competition works, and they trust their ability to pick the finishers and the winners. They are much less sanguine about their ability to predict the outcome of regulatory processes.

Investor responses to the FCC proposal indicate that the form of regulation does matter to the financial community. More specifically, those responses indicate that regulatory processes do influence investor expectations and perceptions of risk and investment growth opportunities. They indicate, in short, that the terms under which they will make capital available to this industry in the future are very sensitive to the performance of regulators in various venues.

» PART II

Perspective of the Service Providers: The Regulatory Risk

4

» Pricing of Local Exchange Carrier Service

Leland W. Schmidt

A way of addressing the issue of local exchange carriers' (LECs') pricing is to imagine the characteristics of an LEC's price structure if there were no regulation at all. What would a telephone company do if it did not have to ask? Which concepts would it use if its only "regulation" of prices were imposed by its customers? It would not be discussing cost-based pricing. One could even imagine the use of a demand curve in setting prices for LEC services. That is not to say the phone company need know nothing about costs. But the costs it would want to understand are long-run and short-run marginal costs that result from specific management decisions.

How could this demand-based concept work with open network architecture (ONA)? To use an analogy, one can buy a Ford by going to a dealer and buying a package—a whole car. Alternatively, one can also go to a parts supplier, buy all of the parts for the car, and pay someone to assemble a Ford car. Of course, not many people would want to construct the whole package from piece parts. But the opportunity is there to buy only what you need, the basic part elements that you want at a compartmentalized price—but the compartmentalized prices do not add up to the price for the whole package.

One could continue to draw out this idea, but even if available competitive alternatives are close to market reality, that recognition is not at all close to political reality. So what does that leave? It leaves pervasively regulated LECs with the problem of defining second- and third-best answers for muddling through a political environment that uses a rearview mirror for defining its policy structure and policy decisions. Even to define an acceptable set of second- and third-best answers for an LEC pricing system requires an implicit set of assumptions relative to the nature and extent of competition in the local telecommunications arena at some future time.

To define that set of assumptions requires some understanding of three factors: (1) future technologies and their costs, (2) the coming set of regulatory and court decisions on competitive issues, and probably most important, (3) the market assessments and decisions by various potential competitors about their business opportunities. Regarding the first point, the evolution of technology useful for performing information-handling and transport functions will inevitably allow ever greater segments of the population to select alternative means of filling their information needs. This will increasingly occur because of, or in spite of, various public policy decisions of the regulators and courts. It is only a question of time.

Second, when faced with open entry questions, the states' regulators, like the federal sector, will opt for more open entry and more competitive supply opportunities. Third, the speed with which competition develops is directly dependent on decisions made by a host of potential market entrants. Market entry decisions by competitive suppliers have not come close to exploiting opportunities available from public policy decisions already made.

Within this context, the issue of appropriate LEC pricing is approached. Assuming erosion of entry barriers, assuming some level of technology evolution that enables new opportunities to be created, and assuming entities willing to enter the fray, the fundamental question is whether LECs' economies of scale and scope in the local arena generally win out over economies of specialization of other potential suppliers.

If LEC economies of scale and scope are improving and are allowed to be made evident in the LECs' prices, LECs will remain the primary suppliers of information transport within their service areas in spite of open entry.

However, LECs could find themselves faced with a broad array of competitors in virtually all important sources of revenue. There are two fundamental reasons why the LECs could find themselves facing broad competition in all its market segments: (1) a lack of sufficient economies of scope and scale relative to economies of specialization of other telecommunications service suppliers; or (2) an inability of the LECs to convince the regulatory process to let them display the economies of scale and scope in their specific service prices.

In the first case, alternative suppliers of specific kinds of services—for example, cellular or integrated shared tenant systems—could erode the revenue base of an LEC faster than the LEC could displace costs, thereby burdening the remaining groups of customers with higher prices. Those or other suppliers again repeat the cycle with new groups of customers. It could be the proverbial slippery scope.

In the second case, the LECs could find themselves in such a situation (erosion of market share and revenues) in spite of their economies of scale and scope. Unless the regulatory process seriously undertakes a reordering—a rebalancing of the sources of LEC revenue to that more appropriate for a competitive environment—continuation of mandated, broadly based "voter" subsidies has a significant possibility of overwhelming LEC economics. But a direct elimination of broadly based voter subsidy is probably too much to hope for—a "first-best" solution in a world of third- and fourth-best realities. But to even deal with third- and fourth-best realistic solutions, one needs to redefine a purpose. LECs and public policymakers do not have a mutual understanding of what each of them wants. Toward what objective are decisions being made? Maintenance of POTS (plain old telephone service) and voter subsidies? Development of a feature-rich network?

An ongoing societal objective, a common purpose, will now be proposed. A fundamental objective of public policy in the United States should be the creation of an efficient and effective telecommunications and information-handling infrastructure. This infrastructure should be one of increasing scope and functionality in order to achieve the broadest connectivity reasonably attainable. If this objective is for the most part accepted, then what remains is to discuss the means by which it can be reasonably achieved.

To achieve this objective, one needs to have an opinion about which of three options for structuring the industry appears most viable:

1. LECs are an anachronism and are not necessary in the longer term.
2. LECs are a necessary, but not an exclusive, element to provide an appropriate local infrastructure.
3. LECs should be the chosen local infrastructure vehicle, and public policy decisions are required to ensure such an outcome.

If one's opinion leads to option 1—LECs are an expendable anachronism—public pricing policy can stay the course. The telephone industry can continue some level of voter subsidies on a POTS network for a long time. But who would continue to invest in such an outcome?

Even the LECs would see that the way to survive would be to invest in things not defined as LEC functions and build a corollary business—in effect, to compete with themselves. It seems that by following option 1, public policy would cause a self-fulfilling prophecy. The LECs are treated as anachronisms, and they would become so owing to lack of new investments and eroding markets. In such a case, who is the next candidate to provide "universal service"?

If one's opinion leads to option 2—LECs are a necessary but not an exclusive provider of a local infrastructure—one would believe that the achievement of an appropriate local infrastructure can be left to a market play-out of a number of suppliers competing for various market segments.

That opinion suggests a continual erosion of regulation of LEC pricing to allow them to react in the market in a manner similar to all other suppliers. Implicit in this option is the necessity for all prices of the LECs to reach their market level. Voter subsidies must erode, and universal service can still be accomplished by subsidies targeted directly to the needy. This subsidy process is best accomplished outside the LEC pricing structure.

If one's opinion leads to option 3—the LECs are anointed as the chosen local infrastructure supplier—one would need a lot of public policy decisions, and fast. The public policy decisions would involve definitions and regulatory oversight/control mechanisms.

Trying to define telecommunications activities that an LEC could perform but would be off-limits to other non-LEC entities, is virtually impossible. Some simple examples: Who is an LEC in an evolved geodesic network? What is a local call? Which information transfer type is defined as telecommunications?

If the definitions are conceived, it is impractical to believe they could remain stable for more than a couple of days. But, dreaming the impossible dream, assume such sustainable definitions are possible. How does one create the regulatory oversight/control mechanisms to ensure that all 100 million end-users and the thousands of service suppliers and manufacturers adhere to the rules? It boggles the mind.

All of this leads back to our original purpose—to explore the future for an LEC's pricing of its services. The best set of assumptions basically emanates from option 2. LECs will be, can be, a reasonably effective vehicle to supply the local infrastructure in a competitive supplier environment.

There is a significant caveat to that statement: it is assumed that the LECs will be allowed by the regulator to alter their sources of revenue to that more appropriate to a competitive environment. By rebalancing LEC sources of revenue, an environment will be created in which LEC economies of scale and scope can be rationally tested against the economies of specialization of alternative suppliers. Out of such an interaction can evolve an effective and efficient local infrastructure system.

For LECs searching for third- and fourth-best solutions, the following set of rules is suggested:

1. No service, in any given situation, can always be defined as competitive. So avoid getting into a situation where one has to "prove" competition to get flexibility.
2. In a multiproduct firm—such as an LEC—at some level LEC services are substitutable. Use that flexibility to achieve revenue source rebalancing.
3. As technology evolves, offering more capability at less cost, more LEC services will have competitive alternatives available. No regulatory control will be necessary.
4. As competitive suppliers make market decisions, more LEC services will have competitive alternatives. No regulatory control will be necessary.
5. In an ONA environment, basic service elements (BSEs) can be used by a competitor to construct any packaged service the LEC can offer. This will not only force revenue source rebalancing but LECs can use it as a mechanism to achieve altered revenue sources.
6. Markets and technology are fluid; current regulation is static. Telephone companies cannot change market forces, and

they cannot change technological progress. So they had better change regulation.

Given sufficient market pricing flexibility, the LEC does possess economies of scale and scope to the degree that it can successfully compete in an open entry environment and fulfill a societal objective of a feature-rich network infrastructure that offers an expanded array of services to a broad customer constituency.

5

» *Local Exchange Carrier Pricing in a Changing Regulatory Environment*

A. Gray Collins, Jr.

Pricing policies in the telecommunications industry have changed as the industry has been faced with increased competition and technological advances. As a result, the industry is currently in a transitional period, moving from rate-of-return regulation to increasing reliance on competitive factors.

Changes in regulation and pricing strategies are necessary if consumers are to realize the benefits that a truly competitive environment can provide. As regulatory restrictions are removed, as they must be, local exchange carriers (LECs) will be able to accelerate their efforts toward pricing and marketing policies that are increasingly responsive to customer needs.

Following is a discussion indicating how the form of regulation affects LEC pricing and marketing behavior. Also included is a framework outlining pricing options that LECs should assess in this increasingly competitive environment. In order to achieve the necessary balance between regulation and reliance on competitive factors, the industry and its regulators must work together to establish a process that will allow LECs to compete while ensuring service quality and universal access to the network.

» *The Rate-of-Return Regulatory Environment*

The traditional form of telephone company regulation has been based on revenue requirements. Rates are set so that the company's revenues will cover its costs, including an authorized rate of return.

Rate changes are granted through general rate increases and specific rate actions. A primary focus of regulatory proceedings is the identification and allocation of costs. Not all costs in the telecommunications industry can be simply allocated to product lines on the basis of direct causation; there is a great deal of shared investment that serves multiple product lines. Much time and effort and many resources are spent on the complex issue of allocating costs, yet there is no lasting resolution. Every time a rate case is filed, the expensive and time-consuming process begins anew.

In addition, rate-of-return regulation has incorporated numerous political and social objectives. Decisions are made that result in cross-subsidies. For example, the rate levels for basic exchange service are not justified from a purely economic viewpoint, but they do accomplish social goals. Economic considerations are given less weight in the telecommunications environment than in other competitive industries.

Cross-subsidies are much less likely to exist when competitive forces are driving prices to costs. To the extent that subsidies exist, they should be more targeted so that they address specific socioeconomic goals. If low-income consumers or specific geographical areas are of concern to regulators and LECs, provisions should be made to assist those consumers or areas. A system that uses averaging to spread the subsidies across all consumers and geographical areas can no longer be maintained. Such an averaging system would disadvantage the LECs by placing on them additional costs that are not being placed on competitors.

As it becomes more difficult to obtain funding for subsidized activities from within the industry, it may be necessary to meet these targeted concerns through nonindustry mechanisms, such as tax-generated sources of funds. As the industry becomes more competitive, this alternative should be evaluated seriously by the industry and its regulators.

Finally, rate-of-return regulation reduces incentives for LECs to be efficient and innovative. Given the freedom to participate in competitive markets, the supplier and the customer can both win: customers get more services for their money, and the firm recognizes an improvement in profitability.

Under rate-of-return regulation, an LEC at its authorized cap cannot keep the additional earnings that cost-reducing innovation or the introduction of new services would provide. Therefore, a firm's incentive to minimize costs and provide new services is greatly reduced or eliminated.

In other words, the earnings of the regulated LECs are limited by their authorized rates of return. An LEC that produces an innovation resulting in significant increases in profitability might be able to retain only the portion associated with the cost of capital. As LECs consider future services, many of which may be less capital-intensive than existing services, they could find their earnings severely limited by rate-of-return regulation. Since smaller amounts of capital would be used, smaller revenue streams would be authorized through the rate-of-return process.

All of these factors point out that rate-of-return regulation has become increasingly burdensome in today's telecommunications environment, which is characterized by extremely rapid technological change and new service offerings. This is clearly illustrated in a recent National Telecommunications and Information Administration (NTIA) study (NTIA 1987) which reports that the cost of rate-of-return regulation to consumers is approximately $1 billion per year. This translates to approximately $8–10 per access line per year.

» *The Transitional Period*

In the last few years, there have been a variety of regulatory reform initiatives, at both the state and federal levels, that differentiate between competitive services and basic services. These initiatives generally retain basic service under close regulation and rely on market forces to influence behavior in the provision of competitive services.

Advances in technology have stimulated this trend, as there are more alternatives for consumers in selected markets. Private networks, microwave, fiber, and increasingly sophisticated terminal equipment are just some of the forces that have fostered this competitive environment. These technological advances are leading regulators and the industry to consider alternatives to rate-of-return regulation.

A major concern of state commissions and the Federal Communications Commission (FCC) is the level of rates for basic residential service. LECs share the concerns of their regulators, and they recognize the need for some form of rate stability that would guarantee rates at current levels or specify the amounts by which rates can change. During the transitional period, rates for basic service will either be sustained or minimally increased. Also, cross-subsidies for basic service are likely to continue to meet the socioeconomic goals of regulators.

LECs are willing to accept the risks of limited pricing flexibility for basic service if they can receive increased flexibility in competitive markets. The objective is to recover the financial shortfall from the provision of basic service by utilizing the freedoms made available in competitive markets.

Why is this risky for the LECs? Because the transitional environment, characterized by a mixture of regulation and competition, limits the LECs' ability to rely on rate cases as a means to recover costs. Innovation and efficiency will be even more critical in order for LECs to provide both basic and competitive services.

Furthermore, LECs will need to become more knowledgeable about their markets—their customer needs, the market positions of their competitors, and the rate and adoption of new technologies. With these expenditures, it is quite possible that revenue streams from competitive services will fall short of covering costs.

However, LECs must accept this challenge and actively participate in this risky game because technology is driving the industry toward increased competition. Technology evolves without regard to regulatory constraints. Competitors of the LECs are able to take advantage of technological innovations, whereas the LECs are currently constrained by regulation.

LECs that are successful in the new environment should not be limited by an authorized rate of return; LECs that are unsuccessful will have to bear the consequences of their failures. The question then becomes, how can LECs manage this transition so as to increase their chances of being viable competitors.

» *LECs in a Competitive Industry*

Firms in competitive industries have the freedom to enter and exit markets, to price at the market level, to alter prices as the environment changes, and to reallocate resources as the market dictates.

Each firm decides appropriate actions through its assessment of conditions in the market. As an example, look at the computer industry, which is somewhat similar to the telephone industry but has more market flexibility. There is little concern about how a firm like IBM establishes prices between competitive products such as mainframes and PCs. This market philosophy needs to be applied to the LECs in order for them to have the flexibility to price their competitive services.

Regulatory bodies have a difficult job. They must determine when competition becomes extensive enough to replace regulation as the best means of protecting the public interest. There is danger in waiting too long before removing regulatory constraints.

The railroad industry, for example, was heavily regulated as it faced increased competition from the trucking industry. Eventually, the advantage belonged to the truckers as railroads coped with outdated regulations that hampered their ability to price according to market forces and to economically deploy new technologies. Many railroads were unable to survive during their wait for regulatory reform. Railroad regulation was relaxed eventually, but for many railroads and their customers, it was too late.

This example is particularly relevant for the telecommunications industry. LECs are facing increasing competition and are hampered by regulatory constraints. The lesson to be learned is clear. Regulators must allow LECs to use the tools they need to compete, when they need them. Otherwise, there is not a fair and level playing field for all participants.

Timeliness is the key. Competitors will seize opportunities that arise in the marketplace. The LEC that is slow in discerning and responding to business opportunities will jeopardize its success in the marketplace.

The method of response will be determined by the LECs' ability to meet their customers' needs; the LEC that is not responsive to changing customer needs (including the implementation of new technology) places itself at a disadvantage.

Competitors' market positions will limit the extent of pricing and marketing options for LECs. An LEC may be unable to establish a desired market position because of a competitor's cost or service advantage. As a result, LECs will be unable to sustain high-margin price levels if a competitor captures the market with a lower price. Therefore, LECs must consider both cost and market dynamics in their competitive strategies.

The LEC will also need to constantly monitor and reevaluate its pricing policies. As market conditions change, the LEC may need to adopt new pricing strategies. Pricing will be the key to success in the new environment. Pricing policies of the future need to emphasize more market and technological factors.

Pricing will have to be more sensitive to the changing needs of the customer. Special incentives could be used to promote new service offerings. Customized services, rates, and billing arrangements are appropriate responses to specific customer requirements.

In addition, pricing will have to be very sensitive to technology changes. It will be necessary to develop pricing policies that incorporate the most cost-effective technology. Cost advantages from the deployment of new technology will be important factors in developing and maintaining the LECs' success in the marketplace.

» *Pricing Options for the Future*

As pricing margins narrow because of technological change and increasing competition, LECs must provide a pricing system that will ensure that costs are covered, while also generating sufficient returns for investors.

Most importantly, the system must also be responsive to customer desires regarding both rate structure and levels, as well as service requirements. A basic tenet of an effective pricing system should be that there is no distinction regarding customer type or information type. LECs should be indifferent about whether the user is a business or other type of customer and should also be indifferent about whether the user is sending voice or data.

The key issue is that a user requires certain capabilities and that rates should be set to recover the appropriate costs of providing service. Who the user is and what information the customer is sending are not relevant if the costs are the same for the transmission.

Another important keystone of an effective pricing system would be the separation of charges for customer access to the network as opposed to charges for transport within the network. Customer access and network transport have different cost and market characteristics, which should be recognized.

A customer access pricing system should primarily reflect the non–traffic-sensitive nature of connecting a customer to the network. Prices should be set to cover the fixed costs of providing customer access to the network. These costs should be recovered through a fixed monthly rate.

Prices for customer access should also be sensitive to the need to reflect market conditions. As new players offer viable alternatives to an LEC's access, the LEC would have to set its prices to meet that competition.

It is possible that the customer access market would change even more radically than the network transport market as various customer access technologies, such as cellular service, evolve.

In contrast to a customer access pricing methodology, a pricing methodology for network transport would be based on distance sensitivity and volume considerations (for example, bandwidth). Of these two, volume will become the more critical pricing component. Deployment of new technologies such as fiber optics will cause the cost predicated on distance sensitivity to become relatively flat, owing to the nature of fiber optics, and therefore should not significantly impact pricing.

Network transport will not be defined by service (for example, voice or data). Rather, network transport is becoming a commodity. As a commodity, users would pay for the basic network transport service and would pay additional charges for optional services or enhancements.

In effect, customers will see a simpler rate structure for network transport due to technological advances. There will be no delineation between business and residence customers; the same rates for network transport should apply to both.

Following are some specific services to illustrate the changes that should occur. First, carrier access charges should decrease as digital equipment becomes more widespread and the cost of network transport declines. A significant portion of LEC revenue received from sale of network transport will be subject to reduced profit margins.

Second, customer access line costs should be covered by fixed monthly rates, while some support mechanisms for social objectives, such as life-line programs, should continue. For example, the installation costs for connecting customers to the network can be substantial. However, a support mechanism could be introduced whereby the LECs can ensure that customers can be connected to the network by providing an option that allows the customer to pay for installation costs over a period of time.

Third, for dedicated types of services, such as private line or special access, there should be greater efforts through contractual arrangements to recover costs over the life of a contract with the customer.

A corollary to distinguishing between usage and customer access would be the elimination of flat-rate service at noncompensatory price levels and the adoption of measured service pricing. Flat-rate pricing is really an averaging system; customers do not pay for the usage costs they have directly caused.

Future changes in usage patterns will make it very difficult to continue providing unlimited usage at a low flat rate. With the advent of new technology, customers are likely to have widely varying usage demands upon the system. There will still be those who have limited use for a phone, but there will be many who will place increasing demand upon the network because of their reliance on the additional capabilities of the network.

Flat-rate levels have been based on average usage patterns. As the amount of overall usage rapidly expands, it is doubtful that there will be enough cost savings from advances in network transport to support continued, low-level flat rates for those making increasingly heavier use of the network.

LECs recognize that many customers assign significant value to a system that provides the ability to make an unlimited number of calls, despite their own personal usage patterns. LECs could meet this market need by offering flat-rate service as a premium service. If customers are willing to pay for the ability to make an unlimited number of calls per month, this pricing option should be made available to them. Rate levels for flat-rate service would reflect the additional value received by the customer. Similarly, option packages could be offered in which the customer specifies the amount of usage per month and rates are set accordingly.

There will be customers who would prefer other types of pricing or billing options. The LEC will need to develop pricing mechanisms that meet these market needs, such as special packaging of calling features or billing arrangements. Again, the particular pricing mechanism will be determined by the needs of the customers and their willingness to pay.

In addition, LECs will need to offer volume discounts to meet the needs of large-volume network users. As previously stated, efficient use of the network will be crucial in a competitive environment. The existence of large-volume users results in increased usage of network capacity. LECs must price efficiently in order to keep large users on their systems. If LECs lose these customers, the fixed network costs are spread over fewer users.

Even in today's environment, competitors view large customers as desirable additions to their systems, and there is intense competition for their business. As a result, larger customers will have more service and pricing options.

Also, large customers have communications needs that can make effective use of many alternative technologies. In fact, some customers currently bypass, and may continue to bypass, the LECs altogether. This base of customers with competitive alternatives will get larger over time as various technologies become more cost-effective.

The quandary for the LECs is how extensively to use the volume discount option to retain customers. Each LEC will have to make judgments regarding the proper structure and level of volume discounts.

Another option to be considered by the LECs is to increase the use of time-of-day pricing. This can be an effective way to control outlays for additional investment. More extensive use of peak/off-peak pricing could provide incentives for customers to change their calling patterns, shifting traffic to time periods when sufficient capacity is available.

This shifting could mitigate the need for investment expenditures that ultimately result in higher rates for network users. Time-of-day pricing has been effectively used in the toll market, providing incentives for customers to alter their calling patterns. There would certainly be some difficult issues to confront before applying time-of-day pricing to local service, but the possible cost savings for customers warrants further examination of this option.

A final point concerns the issue of timely capital recovery. New technology is being rapidly introduced, in effect making existing plant obsolete more quickly than expected. LECs' competitors take advantage of these rapid changes—they use the new technology, which is generally more efficient and less costly. Since these competitors are relatively new players in the marketplace, they have little existing plant that must be recovered.

Existing providers, however, still have plant that is becoming obsolete, and the capital may not be recovered fully, owing to inappropriate depreciation policies. In the future, LEC pricing must reflect shorter useful lives so that the capital expenditures are recovered and future capital will be available for investments needed to provide service to customers.

While shorter useful lives means that capital recovery must occur more quickly, competitive forces will limit the ability of LECs to increase prices. LECs will need to establish depreciation rates that reflect the economic life of the investment needed to provide service.

» *A Cooperative Approach is Needed*

It should be noted that regulators have been responding to changes in the environment; witness the number of regulatory reform plans in place or under discussion at the state and federal levels.

While committed to maintaining service quality, LECs also recognize that regulators do have social and political responsibilities. Certain socioeconomic goals, such as the preservation of universal service, will continue to be important. That is why life-line programs, such as Link-up America and the high cost fund (HCF) have been established to address specific consumer and geographic needs.

In this transitional period, removal of rate-of-return constraints would serve the public interest. LECs would have greater incentives to be innovative and efficient in order to find less costly ways of providing service. A de-emphasis upon rate-of-return regulation would speed introduction of new services and responsive pricing.

What role should regulators play as this new environment unfolds? Regulators must recognize that the competitive environment is here, and that regulatory burdens must be removed where competition now serves as a sufficient regulator. Delaying the removal of these burdens only hampers the ability of the LEC to respond to the dictates of the marketplace and will deny customers the introduction of efficient, low-cost services.

It is imperative that the LECs and regulators work together to establish a new regulatory process that accommodates competitive realities while also maintaining quality of service and universal access to the network.

» *Reference*

National Telecommunications and Information Administration (NTIA). 1987. *NTIA Regulatory Alternatives Report.* Washington, D.C.: U.S. Department of Commerce, National Telecommunications and Information Administration (July): 23.

6

» Cost Allocation Rules for Exchange Carriers for Enhanced and Basic Bottleneck Services

Thomas E. Quaintance

In the network of the future, there will continue to be a very fine line between unregulated enhanced services and the core "bottleneck" services, which the local exchange carriers (LECs) will continue to provide, largely on a monopoly basis.

The exchange carriers have indicated that they will be full participants in the enhanced and information services; they have no intention of being limited to basic POTS (plain old telephone service) communications. If exchange carriers are to be permitted to provide both regulated and unregulated enhanced services without structural separations, safeguards must be developed to protect against monopoly pricing and cross-subsidization.

The key to permitting LEC participation in providing enhanced and information services lies in two critical areas: (1) the identification and separation of common assets and resources that are used for nonregulated operations, and (2) the pricing, unbundling, and tariffing of exchange bottleneck facilities for purchase by multiple enhanced service providers.

The first area is covered by the Federal Communications Commission (FCC) standards for apportioning common costs between regulated and unregulated activities, as well as by the rules for affiliated transactions.

These new rules in Part 64 of the commission's rules are intended to apportion certain common assets (such as central office equipment) and the people who maintain and support them, between regulated network services and enhanced, unregulated services.

If, for example, a large central office switch is used to provide an unregulated voice mailbox service, the switch investment and operating costs would need to be apportioned between regulated network services and enhanced mailbox services.

In its decision in Docket 86–111 (the Part 64 rules), the FCC selected fully allocated costs as the appropriate method for transferring costs between regulated and unregulated businesses of the same firm. That is to say, all direct costs that are identified must carry a porportional share of the overhead costs of the company.

The costing standards attempt to utilize direct assignment of costs as much as possible, followed by attribution of costs based upon usage. Network plant investments are assigned based upon a forward-looking allocator, which is based upon the peak forecasted usage for unregulated services for a three-year period. Investments assigned to unregulated activities cannot be returned to the regulated business without FCC approval, and costs for additional capacity that are to be transferred from the regulated to the unregulated business are at original cost plus return on investment.

All the large LECs, as well as AT&T, were required to file their cost manuals with the FCC for review and approval. The Part 64 allocation rules apply to all LECs, with the exception of the very small "average schedule" companies.

The first area of safeguards was addressed by the FCC in Computer Inquiry III (CI-III), and the second in Docket 86–111 on separation of costs for unregulated activities.

» Regulatory History

In its 15 May 1986 decision in CI-III (CC Docket 85–229), the FCC unveiled a new concept for the telecommunications industry, which it named open network architecture (ONA).

When implemented, ONA will provide a form of equal access for information, enhanced, and value-added carriers to enable them to "plug into" the core telephone network. In exchange for implementing ONA, Bell operating companies (BOCs) will be able to provide enhanced and information-type services without separate subsidiary restrictions. The Phase I order in CI-III only applied to AT&T and the BOCs, and the FCC, on 26 March 1987, issued a further order clarifying that it would not extend ONA to all exchange carriers at that time.

The Phase I order decided several issues. First, AT&T and the BOCs are no longer required to offer enhanced services through a separate corporate subsidiary.

Second, to offer enhanced services generally, the BOCs and AT&T must, on a case-by-case basis, develop and file plans with the FCC for ONAs that provide unbundled basic service elements (BSEs) to others on a tariffed basis and that incorporate comparably efficient interconnection (CEI) technical/operations parameters. Until the network architecture plans are filed, they must obtain authorization for each enhanced service offering, demonstrating that CEI principles are met in allowing other providers of enhanced services access to the underlying basis transport. They must disclose various customer proprietary network information (CPNI) to other enhanced service providers, as well as network information. And they must implement new accounting requirements to prevent improper cost-shifting in the provision of enhanced services.

Third, voice message storage services are enhanced and subject to new requirements for AT&T and BOC provision of enhanced services.

Lastly, states may be preempted by the FCC from regulating AT&T and BOC enhanced service operations. The FCC's attempt to do so, however, has been significantly impeded by the U.S. Supreme Court.

In its supplemental notice on this docket, released 26 March 1987, the commission made several stipulations. CI-III decisions apply to international services. Network channel terminating equipment (NCTE) and multiplexing functions are unregulated. Protocol processing is not regulated. Asynch/X.25 and X.25X.75 protocol conversions can be offered by AT&T and BOCs if CEI is followed. The independent telephone companies are not subject to the same requirements as AT&T and the BOCs. Centralized operations groups

(COGs) should be used when providing basic services to enhanced service vendors. AT&T and the BOCs should use nondisclosure agreements when disseminating information pursuant to make/buy. The BOCs should be required to implement additional restrictions on their use of affiliates' access to CPNI for their enhanced service operations. AT&T and the BOCs should be required to provide information about their networks aggregated from CPNI.

In both of its CI-III decisions, the commission did not describe in significant detail what it meant by ONA, but suggested that ONA services and elements be decided in open forums between the BOCs, manufacturers, user groups, and other carriers. The FCC did, however, provide a few clues. It specified, for example, that carriers must provide unbundled BSEs to others on a regulated, tariffed basis. Some of the other "clues" contained in the order are:

- These components, such as trunk-side interconnections, may utilize subcomponents that themselves are offered on an unbundled basis, such as separate channel signaling and calling signal identification (Paragraph 113).
- As part of any ONA, carriers must provide unbundled basic service "building blocks" (BSEs) to others on a tariffed basis. In essence, competitors will pay only for those service elements that they use in providing enhanced services (Paragraph 214).
- If a carrier's enhanced service utilizes digital transmission, supervisory signaling, calling number identification, and a special alert signal, such as stutter dial tone, CEI for that service should include these basic services as a set unbundled from all other basic service offerings (Paragraph 214).
- Each set of basic services that a carrier incorporates into an enhanced service offering must be available to the public under tariff as a BSE or as a set of BSEs (Paragraph 215).

All BOCs were to file ONA plans with the FCC in early 1988. The FCC is soliciting comments on these plans before reaching a decision on what is expected to be a standard ONA scheme.

Until ONA is implemented, the BOCs may begin to offer enhanced services on an integrated basis provided they provide a CEI to other enhanced service providers on a tariffed basis. There are both technical/operational and pricing principles that the FCC has outlined for CEI.

» *Comparably Efficient Interconnection*

One technical/operational principle is that interconnection opportunities on an equal access basis are ordinarily manifested by nine "parameters": (1) physically and functionally standardized interfaces for transmission, switching, and signaling; (2) unbundled basic services; (3) common basic service rates; (4) common basic service performance characteristics; (5) common installation, maintenance, and repair; (6) common end-user access; (7) common knowledge of availability of basic service features; (8) minimized interconnection cost for competitors; and (9) nondiscrimination among competitors.

Pricing Principles for CEI

All enhanced service providers, including the carrier's enhanced service operation, must pay for the carrier's basic service operations, which includes: rates for transmission facilities based on actual costs of installing and operating such facilities for that provider; an identical basic interconnection charge to recover the costs of the interconnection facilities and the unbundled basic services used by each enhanced service provided by the carrier and others; an equal basic concentration charge to recover the costs of the concentration facilities collocated with the carrier's basic facilities; and standard tariffed rates for specialized communication of signaling services.

Description of ONA

Open network architecture is a concept that has been devised by the FCC as part of the CI-III proceeding; it is intended to promote the widespread use of the core telephone network for multiple use by enhanced and information service (C/I) providers, while permitting the exchange carriers to provide E/I services without separate subsidiary structural separation requirements.

ONA will result in an "unbundling" of the telephone network into regulated, tariffed BSEs that can be integrated by E/I providers into a variety of E/I services. These BSEs can be purchased by E/I providers or exchange carriers and packaged into unregulated E/I services.

ONA is very similar to the concept of access charge tariffs, in which certain access, switching, and transport services have been costed, priced, and tariffed for use by the various interexchange carriers (ICs) to be bundled into their long-distance and private-line services.

In the FCC decision that announced the concept of ONA, the commission left to the exchange carrier industry the definition of BSEs. Although there was some initial discussion at early industry forums about unbundling the telephone network into some very small components, it now appears that the initial set of BSEs will consist of several services or service components familiar today. That is to say, the first BSEs will utilize technology as it currently exists and service elements that are recognizable today.

The first BSEs that have been identified through the ONA forums are:

- Automatic number identification
- Central office announcements
- Answer/disconnect supervision
- Facilities usage information
- Trunk access limitation
- Call detail recording
- Called directory number
- Trunk make busy

Impacts upon Independent Telephone Companies

The FCC decided not to extend ONA requirements to the independent telephone industry at this time. At a minimum, this means that the idependents did not have to file ONA plans with the FCC, as the BOCs had to.

If, however, independent telephone companies intend to provide enhanced or information services in the future, it is highly unlikely that they would be permitted to provide these on an integrated basis without also providing the necessary interfaces with the core network on a tariffed basis.

Assuming that the business decision has been made to offer E/I services as part of a telephone company network, then it is also necessary to devote the resources to developing a comprehensive ONA/CEI plan.

Properly implemented, ONA can provide a positive opportunity to grow network usage and revenues. ONA has the potential for providing a network architecture that provides maximum connectivity to multiple network users. As such, it also has the potential to stimulate new network services and revenue opportunities; reduce uneconomic bypass and maintain end-user ubiquity in the public switched network; support universal service through a better utilized network; increase choices to consumers at competitive prices; and improve economic development, jobs, services, and investments.

BOC Status

In national ONA forums and trade publications, a clear picture is emerging of the progress and status of the BOCs in their approaches to ONA and CEI. Southwestern Bell and Ameritech have made the most progress in developing ONA plans based upon end-user applications. Their preliminary lists of BSEs are oriented specifically toward the uses that an E/I provider would make of the core network in providing E/I services. NYNEX has basically taken the position that all elements required for ONA are currently in their tariffs, and that no further tariffing or unbundling is required. The other four BOCs are somewhere in between these two positions and are obviously still developing their ONA approaches.

Policy Issues

In addition to the costing and pricing concerns, there are several policy issues associated with ONA/CEI that must be addressed. These issues are not unique to ONA/CEI but also spill over into other operational areas. The following are *some* of the policy issues that have been identified to date.

Network Control. To what extent should the network/switching operating systems be opened up to permit customers and E/I service providers to make real-time changes in switching, routing, and service class, or functions, or to have direct access and use of network signaling information?

Colocation. To what extent should the LECs permit E/I service providers, value-added networks, ICs, and so on, to utilize space in LEC central offices?

Geographical Development. To what extent should the LECs develop and deploy ONA/CEI capabilities in their operating areas?

Costing/pricing. What is the appropriate cost context for regulated ONA/CEI BSEs, and what relationship should there be between costs and prices?

Services Unbundling.   To what extent should network services be unbundled to develop ONA/CEI BSEs?

Regulatory Jurisdiction.   Should BSEs be tariffed at the federal level with the FCC, or at the state level with other basic local services?

» *Conclusion*

The purpose of this paper was to provide an overview of the federal regulatory rules that will govern the participation by the LECs in future network services. What remains to be seen is the extent to which exchange carriers are willing to go through the hassle of applying these rules, especially if they are prohibited from directly participating in the attractive information services markets. Judge Harold Greene's recent ruling, which limited the RBOCs' ability to provide information services, may dampen their enthusiasm for proceeding with either ONA or asset separations.

7

» *Preferred*^{sm} *Switched Access: An Overview*

John M. Jensik

» *Introduction and Summary*

On 21 August 1987, GTE filed for a new service offering to be made available to interstate callers in its Florida territory. This service was called Preferred^{sm} Switched Access (PSA). PSA is a new interstate switched access service that provides end-users the option of accessing, on a switched basis, interexchange services that today are reached using dedicated access. PSA was planned to be available within the serving area of the General Telephone Company of Florida (GTFL).

This paper reviews the policy environment that led to the need for offerings like PSA and gives a brief overview of how PSA fits together with the pricing options that long-distance providers have offered. Included in this paper is a description and justification of PSA as a viable alternative to today's interstate access services.

» *Policy Environment*

The current policy environment of the local exchange telephone industry requires that access to the local network by interLATA (local access and transport area) carriers be achieved through what are known as access services.

Access services permit the long-distance carriers to pick up and deliver electronically transmitted information through the facilities of local exchange carriers (LECs). The facilities are made available as dedicated (private-line–like) and switched (addressable offerings) capacity. At this point in time, the dedicated access services may be combined with switched access services and with the functionality of the long-distance companies network to construct a wide variety of service offerings that the long-distance companies make available to their customers.

The prices of these switched and dedicated access services are based on strictly defined, fully allocated cost methodologies. These methodologies assign a significant portion of the common line costs to switched access minutes, while not allowing these costs to be transferred to the dedicated loops associated with providing POTS— the same plain old telephone service that originally made the long-distance business possible.

In addition to inflating the price of a switched access minute by this allocation, a national pooling process has caused all LECs to render a uniform charge (the carrier common line charge [CCLC]) on all interstate switched access minutes (similar to a tax on minutes), which further distorts the relationship of company cost to company price (especially for low-cost companies).

Among the dedicated access services, the uniform "tax" effect that the CCLC causes for switched access services is not present. The outcome is not surprising; if a local exchange company has customers who are heavily reliant on services rendered through the switched access offerings, they soon find that a heavier utilization of dedicated access services lowers their communications budget. This is especially true if the traffic originates or terminates from a small number of locations.

This policy environment has led the long-distance companies to invent offerings that allow their customers to gain access to them in ways that avoid the traditional switched access offerings, therefore avoiding the associated cost burden.

Many of these long-distance companies have logically invented "bulk" long-distance offerings, which give customers the message, "If you can get your long-distance traffic to my switch, we'll deliver it anywhere at a significantly reduced price."

The customer's reaction is, of course, the logical one. "The less I spend in getting my traffic to my long-distance carrier, the lower my communications budget. I think I'll avoid the switched network as much as possible."

The network that was originally designed to carry traffic cheaply for low users is being abandoned by those with high use, and the abandonment will continue as long as the loaded usage price keeps driving high-volume customers away. And as long as the high-volume customers keep leaving, the volume required to make abandonment attractive keeps coming down. This is an especially sorry state of affairs, given that the long-run marginal cost of usage, as shown in what follows, is low.

The answer to this degenerative dilemma is obvious: construct an offering that better reflects the incremental cost of switched usage while deviating in an acceptable fashion from the policies that have established this environment. The answer GTE proposed was PSA. PSA offers to carry traffic to the long-distance company for connection to bulk discounted offerings at a price that is lower than that offered by more traditional switched access services, but only if the customer possesses usage characteristics that would otherwise cause avoidance of the switched network to be the natural economic conclusion. On the other hand, for those customers with heavy traffic from a small number of locations, dedicated access remains the most economic alternative because the cost of any switching is an unnecessary cost for them.

PSA fills a market niche created by current telecommunications pricing policy, while not requiring major modification of that policy. However, since PSA has features not envisioned by Part 69 of the FCC's rules, a petition for waiver was filed concurrently with the proposed tariffs.

» How PSA Works

Recently, interexchange carriers (ICs) have introduced a variety of services that do not include access. These unbundled services, which include AT&T's Megacom, Megacom 800, and SDN, MCI's PRISM, and US Sprint's Ultra WATS, are intended for large business customers. To use one of these services, a customer must arrange to have his traffic delivered to the IC's customer-designated location (CDL), or point of presence (POP). While IC tariffs generally do not specify the type of access the customer should use to reach unbundled services, in practice customers have purchased or leased dedicated access facilities.

The current switched access offerings of LECs are designed to be used by ICs and are neither convenient nor economical for an end-user to buy on an unbundled basis. PSA, in contrast, is designed specifically for this application. It is an optional service, packaged in a form that is convenient for an end-user to purchase. It is designed to maximize the functional benefits of switched access to the end-user. Its volume discounts make it competitive, in terms of price, with alternative access services for end-users with large traffic volumes. It is designed to be compatible with existing unbundled interexchange services. PSA may be used to deliver originating traffic from several customer locations to the CDL of an IC, and to receive terminating 800-type traffic.

A diagram of a PSA service arrangement is shown in Figure 7–1. A subscriber to PSA may designate one or more of his sites to originate and/or receive PSA traffic. (Two customer sites are illustrated on the left side of Figure 7–1). At each site, local connection to PSA service can be achieved either by purchasing local lines or trunks from existing local tariffs or by purchasing special access lines (SALs). In either case, the customer would designate the lines he wishes to be capable of accessing PSA.

Interstate direct-dialed traffic originated on a 1 + basis or on a 10XXX basis (where X is a digit from 1 through 9) from any designated location would be routed over switched network facilities to the unbundled interexchange service selected by the end-user. Terminating 800 traffic from an unbundled interstate 800 offering (such as Megacom 800) would also be delivered on the switched network from the IC's CDL to sites designated by the customer. Access charges for this usage are assessed the PSA rates (described in the next section) and are billed to the customer who orders PSA. The switching and transport functions provided through the PSA tariff are shown between the two dotted lines in Figure 7–1.

PSA service would be available only through designated PSA end-offices. PSA traffic would be delivered to (or from) an IC's CDL,

FIGURE 7-1

≫ *A PSA Service Arrangement.*

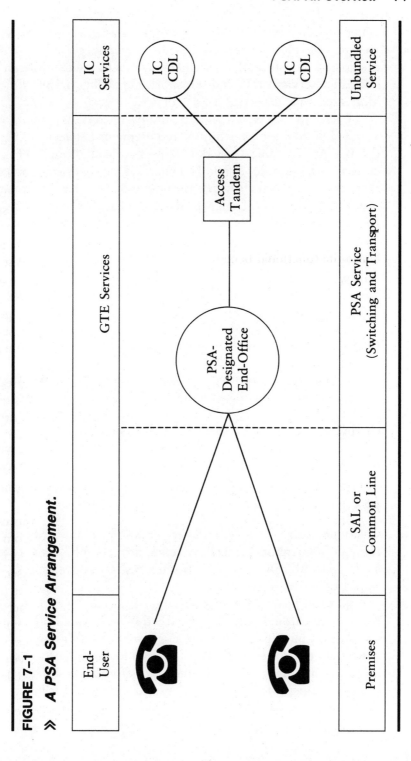

which is served by the GTFL access tandem.[1] If a customer wishes to purchase PSA service and is not currently served by a PSA-designated end-office, the customer may purchase either foreign central office service or special transport, depending on his choice of local access facilities (local lines or special access).

Each end-user would select a presubscribed IC and would work with GTFL to establish trunking requirements between GTFL and the IC. At the same time that PSA is ordered from GTFL, the customer would order unbundled interexchange service from an IC. If the end-user wishes to access the unbundled services of more than one IC, additional CDLs may be designated as part of the end-user's PSA arrangement. Calls may be originated to the additional CDLs on a 10XXX basis, and 800-type calls can be received from those CDLs. The example in Figure 7–1 shows an arrangement in which the end-user has designated two CDLs.

Originating interstate traffic not designated by the end-user for delivery to an unbundled interexchange service would be routed to the presubscribed IC's CDL, as it is today. This traffic is likely to include operator-handled, international, originating 800 and 900 calls. Originating access charges for these minutes would be billed to the IC under the existing access rates. Similarly, terminating interstate traffic, other than unbundled 800-type service, would be carried from the CDL to the end-user, and the appropriate access charges would be billed to the IC, as it is today.

PSA is an interstate access service. All intrastate traffic would be handled as it is today and would not be affected by PSA.

PSA Rate Structure

Customers subscribing to PSA service would be billed directly by the GTE telephone operating company (GTOC) for charges associated with PSA. Charges for local access and usage, if applicable, would be billed separately.

The traffic-sensitive access charges associated with PSA would be assessed through a single rate schedule reflecting both time-of-day and volume discounts. This single schedule would replace the customary traffic-sensitive access charges for central office line termination, end-office switching, intercept, information surcharge, and switched transport. The PSA rate would not vary as a function of the location of the end-user's sites within GTFL's serving area.

PSA would assess a charge per access minute during the day. This charge ($0.02) is less than the sum of customary traffic-sensitive access charges. Evening and night/weekend per minute prices would be discounted 35 and 50 percent, respectively, relative to the day rate.

Customers would be subject to a minimum monthly bill. If the customer's bill, based on all PSA-rated access minutes (originating minutes from all designated sites, plus terminating 800-type minutes), is less than the minimum, that customer would be billed the minimum. A 55 percent discount would apply to the total monthly dollar amount of the customer's bill that exceeds this minimum.

PSA service is designed to allow customers to take advantage of GTFL's inherent network capability to hub traffic at the access tandem. Using PSA service, customers may aggregate traffic from multiple locations to meet the monthly minimum. An initial minimum of $1,500 would apply to every PSA arrangement. The end-user can designate up to three sites from which traffic can be originated and/or to which 800-type traffic can be terminated, via this PSA arrangement. All PSA-rated usage to or from these sites would apply toward the $1,500 minimum. The end-user may also choose to designate additional sites (beyond the initial three) from which PSA calls may be originated. For each additional site the customer designates, the monthly minimum would be increased by $150. The maximum number of sites that can be included in a PSA arrangement is ten (the initial allowance of three, plus seven additional originating sites).

If an end-user designates local lines or trunks as capable of accessing PSA, then a PSA non-traffic-sensitive (NTS) charge of $25.87 per month per customer would be assessed.[2] If a customer designates only SALs as capable of accessing PSA, this charge would not apply.

The computation of PSA charges, including the application of the minimum, is illustrated in Tables 7–1 and 7–2 for two hypothetical customers. (In each case, to simplify the table, the originating and terminating usage to be rated as PSA has been summed for all of the customer's sites.) Table 7–1 depicts a customer who has designated three locations. His total day usage, to and from all three sites, is 100,000 minutes. These minutes are rated at the basic day rate of $0.02 per minute. Similarly, his evening and night/weekend minutes are rated at $0.013 (a 35 percent discount) and $0.01 (a 50 percent discount) per minute, respectively. The total usage charge, before the application of the minimum, is $2,230.

» **Table 7-1.   PSA Customer Bill with Three Locations.**

	Usage	Rate			Bill
Day	100,000	×	$.02	=	$2,000.00
Evening	10,000	×	$.013	=	130.00
Night	10,000	×	$.01	=	100.00
		Total Usage Charge (before minimum and discount)			$2,230.00
		Minimum			1,500.00
		Amount Subject to Discount			730.00
		55% Discount			− 401.50
		Amount After Discount			$328.50
		Usage Charge			$1,828.50
		NTS Charge[a]			25.87
		Please Remit			$1,854.37

[a]In this case, the customer chooses to use common lines from the local tariff to carry PSA traffic.

In this example, since the customer has designated three sites, the initial minimum of $1,500 applies. The total usage charge of $2,230 is greater than the minimum, so a 55 percent discount ($401.50) is applied to the amount, $730 which exceeds $1,500. The usage charge billed to the customer is $1,828.50 (the sum of $1,500 and $328.50, which is the discounted charge for usage over the minimum.)

The customer in Table 7-1 is assumed to have designated one or more local private branch exchange (PBX) trunks as capable of accessing PSA; he, therefore, is assessed the PSA NTS charge of $25.87.

The customer in Table 7-2 has the same total usage as the customer in Table 7-1, but he has chosen to designate three additional sites (for a total of six). His usage total includes originating traffic from the three additional sites. Each additional site increases his minimum by $150, so that his minimum bill is $1,950 ($1,500 + [3 × $150]). The 55 percent discount again applies to the

» **Table 7–2. PSA Customer Bill with Six Locations.**

	Usage	Rate			Bill
Day	100,000	×	$.02	=	$2,000.00
Evening	10,000	×	$.013	=	130.00
Night	10,000	×	$.01	=	100.00
		Total Usage Charge (before minimum and discount)			$2,230.00
		Minimum[a]			1,950.00
		Amount Subject to Discount			280.00
		55% Discount			− 154.00
		Amount After Discount			$126.00
		Usage Charge			$2,076.00
		NTS Charge[b]			-0-
		Please Remit			$2,076.00

[a]Calculation: initial minimum ($1,500) plus per-site additional cost of ($150 × 3) locations.

[b]In this case, the customer chooses only to use SALS from GTOC FCC No. 1.

amount over this minimum, which is $280. The usage charge billed is $2,076. This customer is assumed to carry all of his PSA traffic on SALs, so the PSA NTS charge does not apply.

Basis of Ratemaking

The rates for PSA include several features that are not found in existing switched access tariffs. The GTOCs believe that these rates are appropriate, given the intended application of the service.

The PSA usage rates incorporate nonuniform (volume-discounted) pricing for traffic-sensitive access. These rates were chosen to meet the following objectives:

- To target the service to its intended market of large business users
- To more closely reflect the marginal cost of switched access

- To make PSA competitive in price with alternative access options available to the targeted customers
- To increase the net revenue of GTFL switched access services

The targeting of PSA, in terms of the customer's usage volume, is consistent with the targeting of the unbundled interexchange services with which it is designed to be used. PSA also becomes economically attractive at approximately the same level of usage as the dedicated access alternatives with which it is intended to compete.[3] By targeting the service, GTFL also intended to keep the number of customers choosing PSA within the limits of the systems used to provide the service.

The GTOCs believe that nonuniform pricing can be appropriate for traffic-sensitive switched access because of the cost characteristics of the service. Volume discounts allow large customers to receive a price signal that better reflects marginal cost.

This targeting on the basis of functionality is similar to that established by AT&T with its PRO-America and WATS offerings: a customer who prefers hybrid, WATS-type access and banding would choose WATS, while another customer with the same volume who prefers MTS (message telecommunication service)–like features would choose PRO.

The marginal cost of providing switched access is much less than the average cost allocated to the traffic sensitive (TS) service categories by the commission's rules. This is particularly true for newer technologies—such as digital switching and fiber-optic transport—that involve a significant start-up cost and little additional cost over a very wide range of capacity. All PSA customers would be served by digital end-offices and a digital access tandem. PSA traffic would be transported predominantly over fiber-optic facilities.

If a service with these cost characteristics is priced at average cost, customers receive a distorted price signal. They would purchase less than the optimal amount of the service. In its place, they may substitute another service that is really more costly to provide. For example, a customer may purchase digital special access (DS-1) facilities, even though the economic cost of switched service would be less, because the DS-1 price is less than the average price of switched access. It is not feasible to lower the price for switched service to its marginal cost across the board and still recover the total cost of the service.

Nonuniform pricing allows the carrier to bring the price the customer sees on the margin closer to the marginal cost, reducing the price distortion caused by overhead costs while still recovering the total revenue requirement.

PSA rates have been chosen so that the marginal rate faced by the customer provides a positive contribution above the corresponding economic cost in all three time-of-day rate periods. GTE, in its filing—(GTOC 1987a)—provides a marginal cost study, which the GTOCs believe to be the appropriate methodology for this application, and an average incremental cost study, which demonstrates the sensitivity of the analysis. PSA rates are set at a level higher than either of these measures of cost.

The large customers for whom PSA is intended have a number of access alternatives available to them.[4] In order to reach its target market, therefore, PSA must be priced to meet this competition. In effect, the availability of substitutes makes these customers' demand relatively elastic—compared to that of smaller volume users—with respect to the price of access.[5]

By offering volume discounts to large users, PSA does not create any adverse effect on other switched access customers. The calculations detailed in the transmittal demonstrate that PSA increases the net revenue of GTFL switched access services over test periods of twelve and thirty-six months.

In CC Docket 84–1235, the commission considered a number of mechanisms for ensuring that innovative offerings by a dominant carrier—in that case, specifically, interexchange optional calling plans (OCPs)—were not cross-subsidized by existing services. The commission decided to adopt a net revenue standard. The rationale for this standard is that, if the OCP increased net revenues over a reasonable period, it would not create a revenue shortfall burdening non-OCP customers, and it would not, by definition, be receiving an anticompetitive subsidy from those customers.

In approving the use of a net revenue standard for dominant ICs in CC Docket 84–1235, the commission did not apply a similar test to LEC access services because they were not within the scope of the proceeding. However, the same logic adopted in CC Docket 84–1235 and applied to AT&T's offerings—that a new service that increases net revenue creates no shortfall for any other services to make up—applies equally well to LECs' offerings.

The GTOCs believe that the net revenue test is a tool that can usefully be applied to LEC services. The GTOCs maintain that GTFL faces effective competition in the access market among the large customers for which PSA is designed. Further, the commission accepted this test for AT&T after explicitly assuming, in its Notice of Proposed Rulemaking in CC Docket 84-1235, that AT&T had market power. Therefore, a similar test would be appropriate for a new service offering of a LEC even if the LEC were assumed to have market power in the relevant market.

The net revenue test has proven to be a useful device for regulating the interexchange market. Through the use of this standard, the commission has been able to approve a number of new services, including Pro America I, II, and III and ReadyLine. Customers have benefited from the availability of these new choices. The GTOCs believe that similar benefits can be obtained through the application of this approach to LEC access services.

The PSA rates also incorporate time-of-day discounts. The commission has shown considerable interest in time-of-day pricing and, in fact, required that tapered discount proposals submitted under its guidelines for LEC NTS recovery plans include off-peak discounts. In CC Docket 86–1, the commission considered the application of time-of-day pricing to switched access in general, but decided not to require it.

In their comments in CC Docket 86–1, the GTOCs expressed concern that time-of-day prices for switched access would be distorted if volume discounts were not implemented as well. This would occur because time of day would serve as a proxy for volume, given the difference in calling patterns between business and residence customers (GTOC 1986: 7–9).

This concern is not applicable to PSA because PSA does incorporate volume discounts. Therefore, the GTOCs believe that time-of-day discounts are appropriate for PSA, but remain convinced that such discounts should not be mandated for all switched access services.

While adding new dimensions, such as volume and time of day, the PSA structure has also eliminated some of the dimensions that appear in the existing switched access rates. The rate elements for central office line termination, end-office switching, intercept, information surcharge, and switched transport have been combined into a single usage rate element. This was done because a structure that included all the existing elements, each differentiated by time of day and volume, would have been cumbersome and costly to bill, as well as difficult for the customer to interpret.

Of the existing switched rate elements, the only one that is not currently a simple function of minutes in the tariff (GTOC 1987b) is transport.[6] While it is true that switched transport costs vary with distance, the importance of this relationship is limited for PSA because the GTFL serving area is relatively compact, and because the transport for PSA will be primarily via existing fiber-optic routes. Given a practical limit on the number of variables in the tariff, the GTOCs chose to include the dimensions of volume and time of day, but to make the rate insensitive to the location of the customer within the GTFL serving area.

PSA has been developed to respond to the needs of the market in a timely and cost-effective manner. Existing facilities and systems have been adapted wherever possible. Where new systems have been needed (for certain aspects of billing, for example), they have been developed on a small scale appropriate to the small number of customers in the target market for PSA. The limits established for the number of customer sites in a PSA arrangement have been chosen to ensure that the demand for PSA does not overburden the facilities and the administrative procedures used to provide the service. These parameters have also been tailored to customer usage patterns. Few customers have originating traffic at more than ten locations, and none have significant 800-type traffic at more than three.

Benefits of PSA to End-Users

If PSA became available, all customers would gain an option that has not existed before. Those who select this option would benefit from PSA's combination of switched access functionality, flexibility, and competitive price. Since PSA is optional, each customer can decide whether PSA meets his needs better than other available services. Since the service is designed to appeal to large business users, PSA customers are likely to be sophisticated telecommunications users. Thus, it can be reasonably assumed that these customers would choose PSA only if they would be better off with PSA than with another service offering.

Market research conducted among potential PSA customers in the GTFL market area identified a number of functional differences between PSA and alternative access services that customers perceived as beneficial.

Effect of PSA on Existing Switched Access Services. Because PSA would not change or replace any existing GTFL service, any customer who does not choose to subscribe to PSA would be able to purchase the same service that he would have chosen had PSA not been offered. Because PSA increases the net revenue of GTFL switched access services, the rate of other switched access would not be adversely affected by PSA.

Effect of PSA on ICs.   The introduction of PSA would create a new option for customers to access IC unbundled services. This may have the effect of expanding the market for these services. The flexibility of PSA would make it somewhat easier for customers to change their IC service, or to divide their traffic among ICs. This may have the effect of lessening rigidities in the interexchange market and may create opportunities for smaller ICs to compete for a portion of large customer's business.

PSA would not require any change in the way ICs provide unbundled service today, or in the way they bill for it. Under the proposed PSA service, a customer would be charged separately by GTFL for access and by the IC for unbundled interexchange service. The customer would maintain separate relationshps with the LEC and the IC, unless he wished to designate an agent to represent him.

Effect of PSA on the NECA Carrier Common Line Pool.   Unlike previous LEC proposals for pricing flexibility, PSA is not an alternative NTS cost recovery plan. PSA is essentially a traffic-sensitive access service. However, a small amount of traffic, which would otherwise have paid a CCL charge, is expected to migrate to PSA. A PSA rate element has been developed to compensate the National Exchange Carrier Association (NECA) common line pool for this diversion of traffic.

The development of the PSA charge of $25.87 per month was designed to recover on behalf of the NECA pool the same amount of revenue that would have been paid in CCL charges by the MTS minutes, which are predicted to migrate to PSA. That's $25.87 per customer per month, indeed a low price.

» Postscript

On 2 August 1988, the FCC, after suspending the tariff for almost a year, finally denied PSA. By that time, many of the customers who would have found it competitive had moved to private network solutions. This episode shows the regulatory process falling further out of step with the technological and economic environment. The LECs were not allowed the required pricing flexibility in the face of competition, even when it was demonstrated that the pricing proposal did not harm customers, prevented bypass, and was generally socially desirable. This pricing inflexibility threatens the financial viability of the LEC industry.

» References

GTE Telephone Operating Companies (GTOC). 1986. "Comments of GTE Telephone Companies, Common Carrier Docket 86–1, Phase II" (3 February).

GTE Telephone Operating Companies (GTOC). 1987a. "GTE Telephone Companies Tariff Transmittal No. 292—Preferred Switched Access" (21 August).

GTE Telephone Operating Companies (GTOC). 1987b. "GTE Telephone Operating Companies Tariff FCC No. 1, Facilities for Interstate Access, Ancillary and Miscellaneous Services," as filed with the Federal Communications Commission (1987 revision of 1984 tariff).

» Notes

1. There is one access tandem in GTFL. It is located in the Tampa Central exchange.
2. The development of this NTS charge is described in the filing (GTOC 1987). The local tariff charges for these lines and the appropriate subscriber line charges will be billed separately.
3. PSA is not targeted at every customer with enough volume to consider unbundled service. Within this group, some customers would find PSA more advantageous given their usage characteristics, the number of their locations, and their preference for switched functionality. Other customers with different characteristics and preferences would choose dedicated access.
4. The available alternatives include DS-1 provided by GTFL, line-of-sight and satellite-based microwave systems, and fiber-optic systems. Several ICs have fiber-optic transport facilities in GTFL's serving area. When these carriers bid to establish an unbundled service arrangement for a customer, they frequently have the option of substituting their own transport for all or part of the GTFL facilities that would otherwise be employed.
5. To say that customers' demand is elastic with respect to GTFL access prices because they have substitutes available is, of course, to say that GTFL lacks market power in this segment of the market.
6. The commission's rules do not require that switched transport rates be distance-sensitive. The rates tariffed by Rochester Telephone, for example, do not vary with transport distance.

» PART III

Perspective of the Market Structure: The Monopoly Risk

8

» The Future of Regulation: Lessons from the Recent Past and the Intemperate Present, and Hopes for Future Consensus
Sharon L. Nelson

After the Black Monday stock market crash of 19 October 1987, the financial press was full of apocalyptic analyses, comparing it to the crash of 1929. In a short piece in the *New York Times*, Alvin Toffler (1987) alleged that the references to 1929 were not useful. He believes instead that the financial system is moving beyond the "industrial stage." Reasoning by analogy to the discipline of systems engineering, he likened the 1987 crash to what systems engineers call "hunting behavior." Apparently, this "hunting" occurs when a system becomes unstable and its vibrations amplify beyond acceptable limits.

The current financial system is dependent on computers and communications technologies. Toffler's point is that computer-driven trading programs respond instantaneously to minute variations in prices and volume and amplify their effects. The emergent financial system is a real-time financial system. It is not only faster, it is uniquely diversified and confounded by the complexities of globalization. Despite these observations, Toffler argued against simple-minded deregulation. He said that building a new globally integrated financial system without subjecting it to regulation is like building a super-tanker without airtight compartments. He analogized: "With adequate dividers or safety cells, a big system can survive the breakdown of certain parts. Without them, a single hole in the hull can sink the tanker." He stated that the mindless drive to create a barrier-free financial market is a product of "industrial era" thinking, and cautioned against the almost theological belief in hands-off free-marketism. Toffler's admonitions should be heard by the telecommunications industry.

The attractiveness of the big business market to the telephone industry is well documented. In a recent article, Joe Barrett (1987: 12) looked at how managers link business strategy with information technology. He cited a study by Coopers and Lybrand showing that 46 percent of senior executives thought they would invest in information technology to gain a strategic advantage over their competition within the next two years. However, 65 percent of the same executives went on to say that they believed their technical people do not understand their core business well enough to develop information systems that could truly improve the way people work.

There is a market to be made there. With large American corporations seeking competitive advantage through new high-tech applications, one can understand the almost hysterical mania of telecommunications providers. In this context, all of the rhetoric of the last five years about the need for total deregulation in telecommunications makes sense. In this heady atmosphere, with equality of bargaining power among the players, the celebration of the eighteenth-century ideas of Adam Smith is understandable. The incentives to exploit new technology for use by large corporations have given rise to almost poetic celebrations of the "invisible hand," have added technology to economics in the trickle-down theory, and probably account for the true believers' continued shrill insistence on deregulation.

Regulators understand very well a scenario painted by the proponents of deregulation. In this scenario, if substantial deregulation is not achieved soon, and if the Modified Final Judgment (MFJ) restrictions are not removed, America will slide down a slope of technological obsolescence, increased trade deficits, and general malaise. In the most xenophobic variation of this theme, the United States, in telecommunications at least, becomes a colony of Japan, Korea, and even Singapore. Far-fetched? Probably.

There are other scenarios to be considered as well. The staffs of state commissions and consumer advocates offer equally grim predictions. When they contemplate a largely deregulated telecommunications industry, they envision substantial redistribution of wealth from small customers—who remain captive to public switched networks—to large business users. Since there are no regulators left except those "captured" in the Stiglerian sense, costs will be shifted from the large users to the small users. This would allow providers to maximize their profits while avoiding allegations of predatory pricing. New classes of the information-rich and the information-poor appear. Literacy disappears. The U.S. Postal Service collapses. Dominant carriers take over other embryonic information businesses, ranging from videocassettes to videotext data base suppliers. Newspapers merge with telephone companies to provide electronic classified ads. The Supreme Court decides that telephone companies, like cable TV companies, cannot be subjected to antitrust laws or basic consumer protection regulation because of First Amendment rights. To conjure up an even more Orwellian future: these same companies have an absolute right to invade the privacy of residences—the Supreme Court has held that the distinction between mobile telephones and regular telecommunications technology is no longer valid, and therefore, no citizen has a reasonable expectation of privacy in communications of any sort. Far-fetched? Yes.

Another analyst has taken the "social contract" approach to deregulation and likened it to Texas Instruments offering the nation a similar deal in 1971. In 1971, the deal would have been that the price of calculators for the small business and residential user would remain capped at $100 each—the then prevailing market price. In exchange for that good deal, Texas Instruments would be allowed to divide the geographic market of the United States for "high-end" customers into three parts. West of the Mississippi would go to Texas Instruments, and the eastern part of the country would be divided equally between Hewlett-Packard and IBM. The three providers would agree not to invade each other's territory to get to the high-end markets. Now, jump ahead to 1997: many American corporations have adopted an efficient mix of communications and information-processing. Meanwhile, the captive customer is still paying $100 for a calculator. Some deal. Again, far-fetched? Definitely.

If these visions of the future seem dreadful—and aspects of all of them are dreadful—they are not meant to suggest that regulations or regulators alone could change any outcome. There are limits to the ability of the regulatory or legal system to control how people develop and use tools.

There are historians who, like Hegel, espouse deterministic, single-factor theories of historical change. For example, one prominent French historian attributed the development of medieval European civilization to the fortuitous adoption of one tool—the stirrup. Without it, he reasoned, Byzantium would have triumphed and Charlemagne would never have been. Technological development certainly does contribute to institutional and cultural change. Historians have noted that the power loom ushered in the Industrial Revolution in England. The railroads were the engines, literally and figuratively, of development in the nineteenth-century American West. The better theory, in my view, is that many forces in a society—including regulatory institutions—have an effect on the development of technology.

It is also widely accepted that the public deserves protection from unrestrained private greed as the society moves into future applications of new infrastructure technologies. Just as those subjected to the excesses of the railroad executives of the Gilded Age relied on the trustbusters and created regulatory commissions, many today perceive public utilities commissions (PUCs) as the primary institutions to ensure that the public interest is maintained as we develop the infrastructure for a post-industrial economy.

The society may be on the brink, however, of forging a national consensus about the future of the telecommunications industry and the potential uses of related technology. Paradoxically, this consensus is being formed around the vision expressed by Judge Harold Greene in his latest order: a user-friendly national network where all subscribers to telephone service have the ability to purchase cheap "dumb terminals" that fit into a network capable of providing a wide array of information access and manipulation functions. In Judge Greene's world, while the Bell operating companies (BOCs) are not free from all of the restrictions of the decree, they are free to maximize the potential of their networks. In the process of maximizing the potential for their networks, the holding companies can create a market for new customer premises equipment—and they can market that equipment at unregulated prices.

Judge Greene's vision should challenge us. The judge has not usurped policymaking functions. He has done his job as a federal judge. He continues to interpret the antitrust laws and the decree that was entered into by the parties. His written opinion (*U.S. v. Western Elec-*

tric, 1987a) is a great example of fine judicial analysis, although in one area he has gone out on a limb. The judge has strayed from strict legal analysis in his handling of the information services restriction.

A consensus may be developing among those in industry, entrepreneurs, and regulatory bodies about the validity of Judge Greene's vision. This fragile consensus ought to be encouraged at the national level. The recent joint comments of the US West companies and the American Newspaper Publishers Association before Judge Greene indicate that traditional adversaries can agree (*U.S. v. Western Electric,* 1987b). With a common vision of the future network, or networks, a regulatory structure consistent with that vision should be easier to devise. Unfortunately, the recent rhetoric—the true believer's dedication to the eighteenth-century "invisible hand"—may be blinding the very companies that should accept the challenge of Judge Greene's vision and bring it to life.

Meanwhile, at the state level, experimental approaches to regulatory reform that began in the states of Iowa, Illinois, Vermont, Washington, and New York should provide a basis for optimism. It is, therefore, with a sense of disappointment that regulators at the state level keep hearing calls that often border on fanaticism for a continuation of flash-cut, quick cures. The intemperance of some key players threatens to subvert the progress that has been made toward a rational balance between protecting the public interest and allowing room for the benefits of competition to emerge. On the one hand, emerging competitors argue stridently for what they consider a "level playing field" (an overused metaphor that never provided any real guidance and should be consigned to a rhetorical landfill). To achieve competitive parity, these special interests ask the regulator to ignore the benefits of scale economies and the cost to ratepayers of "leveling" that metaphoric playing field.

At the other extreme, dominant players are demanding freedom to exploit their market power to the fullest. For instance, US West's 1988 legislative proposal for a social contract in the state of Washington represents a radical departure from current ratemaking theory. Prices charged by local exchange carriers (LECs) for still heavily monopolistic services could be totally divorced from anything resembling actual cost. Moreover, it would offer captive customers no real opportunity to protest, while the telephone company raised rates every year for five years.

In their zeal to pursue a corporate goal that is nothing other than "maximum profit with minimum delay," segments of the telephone industry seem to be completely insensitive to the probable response by state regulators and legislatures. The important underlying goals of public interest protection that motivate regulators are not even recognized, much less addressed, in the flood of rhetoric that inundates our hearing records and the media.

A case in point appears in an article by Kaserman and Mayo (1986), two economists who frequently testify for AT&T. The stated purpose of the authors was to debunk the concerns that cause regulators to resist removing regulatory controls. They identified four "ghosts" that allegedly haunt regulators: (1) potential for predatory pricing; (2) the threat to universal service; (3) deaveraging of long-distance rates; and (4) monopoly abuse of rural markets.

The authors then proceed to "exorcise" these ghosts, relying primarily upon the "holy water" known as "fear of potential competition." The authors' analysis was appealing until the following appeared:

> Exorcising this spirit to the satisfaction of regulators, however, is likely to be no easy feat. Difficulties arise from a number of considerations, including the obvious self-interest of regulatory bodies, the equally self-serving claims voiced by the alternative carriers, and the fundamental difficulty of demonstrating the salutary effects of potential competition to non-economists. In essence, one must explain a foreign concept to an unwilling audience in the presence of conflicting claims by adversaries. The disciplining force of unborn competitors is a theoretical concept that is not amenable to either *ex-ante* verification or *ex-post* observation. For nationwide deregulation to occur, a great many regulators must be convinced that this least visible portion of the invisible hand will be sufficiently strong to ward off the quasi-monopoly ghost. Regulators, in general, are not experienced in dealing with the theoretical economic issues involved in the assessment of monopoly power (including, for example, market definition and the analysis of entry barriers). For those accustomed to reviewing reams of accounting figures and cost of capital studies to decide rate cases, the concept of potential competition must appear extremely foreign and somewhat esoteric (Kaserman and Mayo 1986: 84).

It is only human for regulators to react negatively to such condescending and insulting statements. In doing so, it is easy for regulators to forget that there was some validity to the underlying theories that preceded the offensive rhetoric. What is described here is not an isolated phenomenon. It is repeated almost daily, and there is a significant risk that the telecommunications industry, by its underestimation of the regulator's intelligence and by its insensitivity to issues of public policy, may unwittingly retard the development of the types of reforms that can truly meet the industry's needs while protecting the public interest.

The future should hold great promise. The real cost of communications should fall. Opportunities to increase the ability of rural citizens to gain access to information services now available only in the urban centers should abound. The increased use of the network should benefit everyone. Any necessary price increases or cost shifts should be more than offset by improvements in the value of the service. But can this potential be achieved without the support and cooperation of state legislators, governors, and the PUCs? No.

The telecommunications industry now prides itself on marketing. In a competitive world, marketing is a fundamental skill. It is time for the players to begin to market their ideas to the regulatory community and to use a soft-sell approach. Regulators do not respond to scare tactics. They do not buy snake oil. If the industry has a good idea (that is, an idea that has tangible benefits to the public without an unacceptable windfall to the "salesperson"), then it should give regulators the facts, not the rhetoric, and the idea will sell itself. The regulatory community may not be populated with Ph.D.s, but regulators understand the business world, understand the political realities with which they live, and have access to talented support staff to analyze technical issues. They will recognize a good idea when it is proposed.

The collective spirits of regulation are not yet exorcised. I believe that the intemperance of the present is something like adolescence—an exciting, exasperating, frustrating, unproductive, and yet necessary stage in the maturation of the relationship betweeen the telephone industry and its regulators. Perhaps this stage, like puberty, will soon pass, leaving everyone better off for having gone through the experience. The regulators will have learned that the industry understands its marketplace, its technologies, and the potential for bringing the two together for the mutual benefit of the stockholder and the ratepayer. At the same time, the telephone industry will have learned that regulators are uniquely capable of monitoring the public pulse so as to maintain that degree of regulatory control necessary to assure the public that its interests are being protected. One telephone company president confided that he thought his company needed the continued umbrella of the commission to accomplish what he needed to do: raise rates because of federal-state cost-shifting. He was reminded that regulators are good flak-catchers. More importantly and less cynically, the industry needs a good referee to resolve intra-industry disputes. No deregulator has suggested that superior court judges in the states take over the regulatory role in settling intra-industry disputes.

In conclusion, notwithstanding the opinions expressed in the Huber Report (1987), public interest regulation is not "moribund." It is, rather, in a state of transition. If the telephone industry representatives will recognize this and devote their energies to the forging of improved relations with regulators, they will do more for the cause of rational regulatory reform than any number of zealous denunciations of regulation and regulators could ever accomplish.

» *References*

Barrett, Joe. 1987. "Telecom and the Corporate Plan." *Procom Enterprises Magazine* 1, no. 4 (August): 12–14 and 58–59.

Huber, Peter W. 1987. *The Geodesic Network: 1987 Report on Competition in the Telephone Industry* (Huber Report). Washington, D.C.: U.S. Department of Justice, Antitrust Division.

Kaserman, David L., and John W. Mayo. 1986. "The Ghosts of Deregulated Telecommunications: An Essay by Exorcists." *Journal of Policy Analysis and Management* 6, no. 1 (Fall): 84–92.

Toffler, Alvin. 1987. "A Post-Panic System." *New York Times* (24 October): sect. 3, p. 8.

United States v. Western Electric, 673 F. Supp. 525, 540 (D.D.C. 1987a).

United States v. Western Electric et al. 1987b. U.S. District Court for the District of Columbia, Civil Action No. 82–0192. "Proposed Orders and Explanatory Memoranda of the American Newspaper Publishers Association (ANPA) and U.S. West, Inc." (filed 15 October).

9

» Telecommunications: Issues Facing the Bush Administration

Stuart Whitaker and *Alan Pearce*

This paper is intended to provide information and analysis of national-level government activities—specifically, at the Federal Communication Commission (FCC), the U.S. District Court, and in the U.S. Congress—that will affect the future of local exchange service.

» Federal Communications Commission

After waiting over six months in 1987 to fill a single vacant commissioner's slot, President Reagan found himself with a second vacant slot. He nominated Susan Wing, a partner with the Washington law firm of Hogan & Hartson, and Bradley Holmes, Chief of the Policy and Rules Division of the Mass Media Bureau at the FCC. Senate Democrats initially refused to schedule confirmation hearings, in recognition of the fact that a Democratic victory in the 1988 presidential election would afford them the opportunity to select four new commissioners, including a chairman.

Now the hierarchy at the FCC will be changed dramatically. The new Bush administration has plenty of flexibility at the FCC precisely because it is able to name four out of the five commissioners, including the chairman. Only Commissioner Jim Quello, a Michigan Democrat who is in his early seventies, is likely to remain. His five-year term does not expire until June 1992. Because Reagan administration appointees at the FCC were prone to the criticism that they ignored the Communications Act of 1934 and the Code of Federal Regulations that stems from it, the Bush administration will appoint individuals committed to a stricter interpretation.

Once all of these presidential appointments have been made at the FCC and confirmed by the Senate, the new chairperson and the other commissioners will begin to bring in new top staffers; for example, a new chief of the Common Carrier Bureau, a new chief of staff, a new chief of the Office of Plans and Policy, and so on. Perhaps as many as twenty new senior staff appointments will be made to promote President Bush's policymaking initiatives.

The new FCC will attempt to resolve some of the major policy issues that had been "on hold" because of election-year politics. Included among these issues are:

- *Price caps.*   Comments have come in for the proceeding on alternatives to rate-of-return regulation—otherwise known as price caps—and the staff has been sifting through these comments to prepare an agenda item as early as possible. This rule-making will be stalled, however, until the new administration has an opportunity to fully evaluate the issues. The administration must decide whether to press ahead with an extremely controversial policy that, if adopted, will most certainly be appealed to the D.C. Circuit Court of Appeals, and perhaps even to the Supreme Court.
- *Telco-cable and TV network-cable cross ownership.*   Two separate rule-makings, launched in July and August of 1988, threaten to change the structure of the telecommunications/information industry. Before telephone companies can be permitted to own cable systems in their own operating territories, permission of the U.S. Congress is necessary. Both the TV networks and the telephone companies have been kept out of the cable TV business since an FCC rule-making enunciated in 1970. These cable TV ownership issues will not be resolved until the spring of 1989 at the earliest.

- *Computer III and the approval of the BOCs' ONA plans.* The information age is being held in check because of slow progress in approving the Bell operating companies' (BOC) open network architecture (ONA) plans, which were submitted to the FCC in February 1988. Currently, it seems that the earliest these plans will be approved by the commission is spring 1989; approval could come as late as fall 1989. Whichever is the case, the new administration may want to slow the process down, since Computer Inquiry III was the brainchild of the two previous commission chairpeople. The administration may ask the FCC to take a fresh look at the policy.

- *The 800 data base.* The FCC has been struggling to formulate an 800–data base policy since early 1986, with little or no success. At issue is how to legally remove the 800–data base monopoly from AT&T so that it becomes available to a wider array of suppliers—that is, the other interexchange carriers (ICs), such as MCI, U.S. Sprint, and the independent telephone companies.

- *Strategic tariffs.* The most important of the so-called strategic tariffs—special prices offered to major telecommunications users by AT&T and some of the BOCs—are Tariffs 12 and 15, both submitted by AT&T. Tariff 12 was developed by AT&T to serve only major customers—those prepared to ante up $3 million a month or more, such as DuPont, the Department of Defense, Ford, General Electric, and American Express. Tariff 15 is designed to permit AT&T to make a competitive response to price-cutting by MCI and other long-haul carriers. Because AT&T is classified as a dominant carrier and is therefore rigidly price- and profit-regulated by the FCC, it lacks flexibility. Tariff 15 was specifically designed to meet the needs of Holiday Inns, Inc., of Memphis, Tennessee, which was threatening to go with MCI. AT&T's strategic tariffs have resulted in much controversy among FCC staff, and even among the commissioners themselves. The FCC will continue to struggle with pricing issues for several years to come. Ultimately, the FCC will have to resolve them.

» Department of Justice and the District Court

There will be major changes in the management and policies of the Department of Justice (DOJ). For a start, there will not be the vigorous effort to dump the Modified Final Judgment (MFJ) that existed in the previous administration. Under the Reagan administration, much of the effort to dump the MFJ was spearheaded by Attorney General Edwin Meese. Allegations of favoritism were made concerning Meese, who had a close relationship with his predecessor, William French Smith, who in turn maintained a relationship with another former Reagan administration official, Judge William Clark of California. Smith and Clark, who represented Pacific Telesis, were said to be the architects of the push for early relief for the BOCs from the MFJ. This powerful policy push resulted not only in favorable DOJ recommendations to Judge Harold Greene in February 1987, but also resulted in the Dole bill, which was designed to strip all policymaking power from Judge Greene and to give it to the Reagan-controlled FCC.

Both policymaking initiatives failed. Judge Greene totally ignored the DOJ recommendations, and the Dole bill died at the end of the 99th Congress. The Republican majority in the Senate also ended at the same time, and since then no one—not even Senator Robert Dole—has seriously sought to raise the issue of MFJ relief for the BOCs. The Bush administration will proceed slowly, if at all, on BOC MFJ relief. The DOJ may be upgraded—particularly the antitrust division, headed by Richard Rule.

The Bush administration will refuse to interfere with the preparation of the second triennial report to Judge Greene. While he is scheduled to receive this report on 1 January 1990, it is widely believed that a delay beyond that date is virtually inevitable. However, only Judge Greene himself can grant the DOJ a delay, and he may insist that the second report be delivered on schedule. Before taking any policy positions of its own, the Bush administration is likely to wait to see first what is recommended to Judge Greene—perhaps some time in 1991—and second, how the judge reacts.

The result will be not much change in the line-of-business restrictions for the BOCs until mid to late 1991, and even then the relief will be limited, focusing on equipment manufacturing and some slight flexibility in the area of information services.

» *Congress*

If the 101st Congress is anything like the 100th, it will be pro-regulatory as opposed to deregulatory—and the President may go along with the flow to maintain an effective relationship with the

Congress. Antitrust enforcement is likely to be given a big boost by the chairpeople of the two congressional committees with primary involvement in this industry. Congressman Jack Brooks takes over the House Judiciary Committee at the beginning of the 101st session, and Senator Howard Metzenbaum, who has doggedly opposed MFJ relaxation, will remain as head of the Senate Antitrust Subcommittee. Senator Joseph Biden, Chairman of the Senate Judiciary Committee, is expected to be tough on the BOCs and favors MFJ enforcement. All three legislators have supported Judge Greene in the past, as has Greene's George Washington University Law School classmate, Senator Daniel Inouye, who plans to stay on as Chairman of the Senate Subcommittee on Communications.

The question of legislation designed to free the BOCs from the line-of-business restrictions in the MFJ will be left until after the triennial report has been delivered and commented upon and a decision has been rendered by the judge. Even then, passage of a bill by both the House and Senate is highly questionable. In 1988, House Energy and Commerce Committee Chairman John Dingell introduced Resolution No. H.R. 339, which proposes to relax the MFJ rules on the BOCs so that they can provide content-based information services and manufacture telecommunications equipment.

This resolution received 81 cosponsors, 43 Democrats and 38 Republicans. But before the BOCs begin to celebrate, they must remember that 1988 was an election year and many of these "supporters" were merely repaying the BOCs for political action committee (PAC) contributions to their campaigns. The members of Congress have probably all forgotten about H.R. 339. There is also little or no enthusiasm on Capitol Hill for amending or rewriting the Communications Act of 1934, and this would have to be done if any of the major players in the telecommunications/information industry are to receive any significant regulatory relief over the next four years.

» *How It All Adds Up*

Tough regulatory times are ahead for the local exchange companies in the telecommunications/information industry. An active regulatory regime will continue and may be enhanced in the next four years or so. The same fate awaits AT&T, which might explain why it is bombarding the FCC with a spate of regulatory relief proposals, particularly concerning its pricing—or tariffing—strategies and its profit-making abilities. Currently, the FCC and the states regulate both AT&T's prices and profits. AT&T has been awaiting some form of deregulation since 1984, that is, since the breakup of the old Bell system. It will have to wait at least five years—or more!

Eventually, however, the line-of-business restrictions imposed upon the BOCs by the MFJ will be lifted, cautiously at first, and then perhaps more rapidly in the mid-1990s. Even so, a stringent regulatory regime will continue to be imposed upon the BOCs as they enter new areas—equipment manufacturing, information services, and so on. It is unlikely that the BOCs will be permitted to offer long-haul services, largely because of the vulnerability of AT&T to long-haul competition from the BOCs, which have a strong territorial advantage in each of their seven regions. In addition, the BOCs still control the so-called local exchange bottleneck, that is, they have access to almost every American household and office. AT&T will, however, be permitted to enter the electronic publishing business sometime after 1 January 1991, when the MFJ prohibition expires, but here again there will be a regulatory regime imposed, most probably by the FCC.

The structure of the U.S. telecommunications/information industry will change over the next several years, owing in part to FCC policymaking initiatives that will permit the TV networks, and perhaps even the telephone companies, to buy into the cable TV business. Other factors influencing industry structure will be evolving competition in long-haul services and information age services. If the long-haul business becomes too tough for the existing carriers, then there could be mergers and acquisitions. If the going gets really tough, policymakers may be forced to permit the BOCs to buy some of the troubled long-haul companies. As information services begin to proliferate, there will be the inevitable shakeout in the industry, and public policy may permit or even encourage some merging.

Whatever happens with the new Bush administration and future administrations, the telecommunications/information industry will continue to be regulated by the FCC, aided and abetted by the vast majority of the states, and will continue to be subjected to MFJ enforcement from the DOJ and Judge Greene.

Finally, Capitol Hill and the White House will continue to study, analyze, discuss, criticize, and cajole. This is so because the industry has become so important, as society careens toward the third millennium, that the politicians will be unable to leave it alone.

10 » *Drivers of Change In Regulation*

Thomas Chema

In light of the vast changes sweeping the telecommunications industry, this paper's theme would seem to require some rather heavy-duty crystal ball–gazing into the regulatory future of the telecommunications industry in this country. While prognostications of this sort have always produced results about as reliable as reading tea leaves or visiting Gypsy fortune-tellers, this paper nonetheless will assess quite seriously future trends in state regulation of local exchange providers of telecommunications services. I will endeavor at all times to stimulate constructive thought—rather than simply rehash old issues and problems. This assessment reflects my experience in the policymaking realm of state government, in the chairmanship of a state regulatory agency, and training in the legal profession. It will also reflect my intense interest in the formulation and implementation of sound telecommunications policy, at both the state and federal levels.

Before embarking on this journey into the future, it seems worthwhile to spend a few moments recapping briefly what has transpired in the recent regulatory past relative to telecommunications, and then to make another quick stop along the way to describe the regulatory conditions that appear to be confounding the telephone industry at this point in time. State regulators definitely view themselves as significant drivers of change in the utility matters for which they have statutory oversight, but they are certainly not the only game in town. Thus, the forthcoming observations and analysis should in no way suggest that the corporate planners, managers, and researchers of the telephone industry should take the advice of the Greyhound Bus ads and leave all of the driving to the state regulators. Clearly, telecommunications is an industry still in transition. A number of factors have combined to create rapid shifts in the nature and structure of the telecommunications industry, both at home and abroad. While many analysts lament the confusion that seems to have been precipitated by these changes, a number of broad themes continue to emerge.

First, one cannot help but notice the almost frantic pace at which changes in the telephone industry are taking place. Over the last twenty years, telecommunications has been transformed from a monopoly-based, wire-line network designed to serve basic voice and some data needs, to a combination of monopoly and competitive networks that use wire, microwave, satellite, computer, and other technologies to serve a broad range of information-processing and transmission needs. While there may be some dispute as to whether this is truly an information age, it is at least clear that the industrial age society is changing markedly as a result of this new, advanced technology.

The communications network traditionally has been the foundation upon which society builds new services, products, and market applications. The rise of the personal computer, computer-aided design, robotics, and other technologies indicates that large segments of business will increasingly rely upon communications to replace labor, improve productivity, and increase profitability. Ever growing numbers of workers seem to be concentrated in service and information-processing sectors, while fewer workers labor in industrial settings.

As information technology has changed, so too has the corporate structure by which telecommunications services are delivered. The prime catalyst, or driver, of this change was the 1984 divestiture of AT&T. As a result of this divestiture, seven new regional Bell operating companies (RBOCs) were created. These companies have exhibited an attitude toward telecommunications that differs somewhat from the typical public utility concept of the past. Essentially, these regional holding companies have shown a desire to diversify into new markets as quickly as the law will allow. At the same time, financial pressure on the operational aspects of many local independent telephone companies is increasing. The high cost of serving certain rural areas and other factors are combining to drive some small companies out of the market and to make others vulnerable to takeover. Many analysts foresee a substantial concentration in ownership in telecommunications over the next decade. Meanwhile, in the terminal equipment, long-distance, and specialized service markets, there has been an obvious explosion of new products.

New corporate giants such as MCI have developed over the years, and old corporate giants such as IBM are aggressively entering telecommunications markets. Various acquisition transactions are now taking place more frequently, and new partnership agreements are being entered into by these companies on an almost daily basis. Such alliances may be the harbinger of a set of mergers, on an even grander scale, that could ultimately lead to the creation of large computer-telecommunications (or "compucations") conglomerates.

While many aspects of the communications industry have changed drastically, the ability to communicate has steadfastly remained a basic human characteristic and need. Indeed, as society has become more complex and dependent upon sophisticated telecommunications gear, the need for universal, interactive communications service is growing by leaps and bounds. Fire and police dispatch are becoming more dependent upon telecommunications through sophisticated 911 emergency number systems. Social services and community resources are now often dispensed through telecommunications networks. Increasingly, banking and other financial services are relying on computer and telecommunications technologies to reach customers and open new markets.

As society becomes ever more dependent upon telecommunications and information-processing technologies, the need for all segments of the population to have access to telecommunications service, or risk being written out of society, becomes even more clear. While consensus has begun to form on some national telecommunications issues, many critical issues remain controversial and are as yet largely unresolved. Still needed, for example, is a more workable definition of what constitutes "basic service" in markets that traditionally have been served by local exchange carriers (LECs). Should a high-tech network of the future be encouraged? Or should we strive to maintain only basic voice-grade local networks? Should regulators permit newer technologies, such as cellular mobile, to totally replace existing wire-line telephone networks?

While some of the information needed to address these issues is technical and economic, major policy determinations and value judgments must also be made. For example, as patterns of owner-ship in telecommunications change in this country, there continues to be controversy over what the corporate structure of the telephone industry should look like in the future. Should state and federal policymakers encourage, for instance, the continuation of a network of small independent companies serving rural America? Or should policymakers allow concentration of the industry in the hands of just a few large holding companies? Should policymakers experi-ment with diversity of ownership options to a greater degree than they have in the past? In general, the issue boils down to whether the country will be best served by an oligopoly of large, well-financed, technically sophisticated corporations with potential economies of scale and scope in the world markets, or whether policymakers should promote more diversity in the ownership of communications resources, even at the possible expense of network sophistication and trade imbalance. And the primary considerations in making this judgement, it is emphasized, are policy oriented and value-laden.

Regardless of the policy direction ultimately chosen for the future of local exchange service, diversity of ownership supply, and national network design and functionality, the most critical regulatory issue con-fronting us at this time concerns the type of incentives that are necessary and appropriate to ensure the continued high quality and wide availability of telecommunications services to the people of this nation.

Without the addition of better, more direct and more powerful incentives, the United States will never accomplish one of its most fundamental, underlying societal objectives: to create an effective, yet efficient, telecommunications and information-handling in-frastructure, of increasing scope and functionality, that will be capable of achieving the broadest connectivity reasonably attainable. Such incentives should be positive rather than negative. It would be preferable, therefore, to work on developing a system of mechanisms to reward providers of telecommunications services for desired behavior instead of perpetrating a regulatory mind-set that seems to focus only on finding ways to prohibit certain activities and to extend policing and enforcement processes even further.

Positive incentives can be built into the most traditional or regulatory settings, albeit with more effort and more difficulty. And while mixed incentives are always possible, negative incentives should not be explicitly built into the regulatory scheme, but only brought to bear under conditions of extreme urgency, when desired behavioral objectives clearly have not materialized.

Now that the philosophical basis for the type of telecommunications regulation that needs to be embraced at every jurisdictional level in this country has been built, the balance of this paper shall detail the very real benefits to be derived from such a system, the ancillary equipment needed for the system to work, and what all of this could mean to the future of the local exchange telephone industry in terms of continued service vitality, financial viability, and progress.

Quite simply, local exchange service providers must be encouraged in a real monetary sense to develop new and additional uses for their local exchange networks. At the same time they need to be rewarded for finding ways to hold down and even reduce the operating costs of these networks. Additionally, the industry needs to be challenged to explore the development of new service offerings that utilize the local loop, but do so without the need for expensive new equipment.

Clearly, the regulatory goal is to seek for the local loop the same benefits that helped reduce costs on the interstate network side: increased traffic volumes, new and enhanced services, and competition where feasible. For new, discretionary, promotional, or luxury services that utilize the local loop, flexible and market pricing should be permitted so long as appropriate cost standards are met and some reasonable measure of return is provided. While the profits from these new services should be divided in some rational manner between both stockholders and ratepayers, the companies' share should certainly be proportionate to the level of risk associated with these new ventures.

This same type of regulatory reward system should be at play in diversification enterprises proposed by local exchange companies. Instead of looking skeptically at all forms of activities, the regulatory community should exhibit a friendlier attitude toward those ventures that have the potential for generating even more revenues from the local loop. There must be a trade-off here, of course. In return for its portion of new profit and the chance to explore additional avenues of business opportunity, the exchange company must agree not to charge ratepayers for new ventures that are ill-advised and big money losers. These economics of joint development—negotiated on a company-by-company basis—could continue as long as traffic growth and demand for service contribute revenue. Should demand stabilize, the remaining local loop investment will have been reduced to a level at which local rate increases will be either small or unnecessary.

Under the incentive regulatory plan just outlined, it is possible for almost everyone to win something. Local telephone subscribers would get real and not artificially contrived rate stability; utility shareholders would get increased and at least partially deregulated profits; and users' demands and needs would be better met with new services at market prices.

11

» "New" Market Pricing Proposals for Telephone Exchange Services: A Critical Appraisal

John W. Wilson

From a consumer perspective, there is little to commend either in today's "new" pricing ideas or in the pricing policies implemented in recent years for telephone exchange service. There is also nothing really new in today's alternative pricing concepts. They have undergone some significant repackaging, to be sure, but the contents of today's proposals do not differ materially from the strategies of the past.

Rhetorical emphasis has apparently shifted from "cost-based" rates, which were in vogue a short time ago, to "market-based" prices today. But exchange service rates were never really cost-based in any objective sense. As value-of-service pricing lost favor in the 1970s, telephone companies and sympathetic regulators who were anxious to capture the maximum contribution toward system costs from captive markets, quickly declared that local access (loop) costs were properly chargeable as a fixed cost component to basic network service. That arbitrary rationalization, which had nothing to do with economic cost causation, meshed nicely with marketing objectives that could be best pursued unencumbered by any need to recover basic system costs from competitive services. The practice of "reckoning" costs to support pricing objectives became the hallmark of the telephone utility rate filings in the early 1980s. The fact that carts were pulling horses did not bother riders who were happy with the destination.

This practice highlights the fact that it is exceedingly difficult to approach questions of economic policy with complete objectivity, and that the practitioners who are called upon to address economic policy matters are typically not "open-minded." They bring to policy-making minds filled with experiences and preconceptions and a resulting point of view.

The point of view of those advocating telephone service price deregulation and the elimination of the Modified Final Judgment (MFJ) restrictions in the telephone industry is clear. It is reflected not so much in any particular thread of bias running through their policy positions, but rather in the general pattern of the whole fabric. These policy advocates view free-market dominance by large integrated firms with equanimity. They have faith in an economy of large and generally unconstrained private enterprise, and their policy positions reflect that faith. Their solicitude for big business outweighs their fear of monopoly.

A notable example of such advocacy was the telephone industry's attack in October of 1987 on Judge Harold Greene's decision not to lift the MFJ restrictions on regional Bell operating company (RBOC) activity. Anxious for the opportunity to exercise their considerable monopoly power under a free-market banner, the RBOCs were joined by representatives from the Federal Communications Commission (FCC), the Justice Department, and the National Telecommunications and Information Administration (NTIA), who "bashed" Judge Greene for everything from usurping a regulatory role to undermining Reagan administration efforts to reduce the trade deficit.

As the political economy continues to drift in a sea of free-market rhetoric that emasculates Judge Greene's economic and legal concepts of competition with a simplistic, laissez-faire ideology, dominant exchange carriers may, for a time, no longer be obliged to cloak their pricing proposals in a cost fabric. Today's new pricing initiatives are comparatively straightforward proposals to deregulate. While a regulatory label is retained—such as "social contract regulation," or "price cap regulation," presumably in deference to perceived statutory obligations—the prescription is clearly deregulation, with little pretense about any relationship whatsoever between costs and prices.

The largely opportunistic nature of such appeals for public policies to either enhance or constrain market forces is well illustrated by the experience in natural gas markets during the past decade. One of the ironies in the recent roller coaster ride of world energy prices was the public policy turnabout by U.S. gas producers and pipelines in the 1980s, when their fortunes reversed as the OPEC cartel collapsed and prices went into a free-fall. While prices were rising, oil and gas barons in Louisiana, Texas, and Oklahoma sported bumper stickers reading, "Drive a Caddy—Freeze a Yankee." Their lobbyists descended on legislators and regulators, urging them to let the competitive pricing forces of the "marketplace" prevail. Then, when world energy prices collapsed and demands shifted, leaving gas pipelines with massive "competitively" negotiated take-or-pay obligations and gas producers without sufficient markets to support their enterprises, the industry's economic policy imperatives reversed. As gas demand dropped and prices plummeted, unemployment in Texas, Oklahoma, and Louisiana soared, banks failed, and drilling companies and wealthy speculators became insolvent. Gas producers and pipelines responded by seeking government protection and regulatory bailouts to preserve their "critical domestic industry."

To compound the irony, just as the oil industry's most conservative and vigorous capitalists started to exhibit socialist tendencies, a new breed of capitalists, selling competitive reorganization, began to emerge in the electric utility industry—which, naturally, had been the staunchest stronghold of state protectionism from market forces during the oil industry's heyday. So too has the telecommunications industry plied a vacillating course in recent decades between the advocacy of regulatory protectionism or laissez-faire, as circumstances affecting its special interests have changed.

Such apparently inconsistent temperaments are, of course, easily reconciled. Short-term self-interest lies behind the advocacy of both more and less reliance on market forces. Ideological rhetoric gets thrown in simply to sell the current scheme. Because logic counts for little when self-interest and ideology converge, it is entirely consistent for the telecommunications industry to call for more regulation in circumstances when competitive market realities are harsh, and to call for free markets when regulation threatens to constrain potential profits. Whatever the professed rationale, proposed changes in the regulatory status quo generally seek to increase the proponent's relative wealth. If it is currently beneficial to let "market forces" reign, deregulation will be demanded; if gain lies in protectionism, that will be sought. In today's political economy, a "new social contract" pricing package wrapped in market rhetoric and decorated with ribbons of consumer protection seems irresistible. In any case, whenever an industry asks for more or less regulation, consumers should reach for their wallets and hold on tightly.

This is not to say that consumer interests are indifferent in the debate over competition and regulation—although they do tend to be less organized and therefore sometimes less effective politically than other special interests in advancing their objectives. Setting aside the obvious concerns that an economist (or a judge) must have when ideologies confuse simple market freedom (such as freedom for a monopoly to exploit its market power) with effective competition, it is clear that although competition and regulation both promise consumer protection, in the real world it is often the case that neither accomplishes this very well alone.

As every economist knows, perfect competition is a textbook abstraction whose major advantage is pedagogical. Real world competition, like real world regulation, is always imperfect. Fortunately, competition and regulation are not mutually exclusive, and consumers can be well served in real world settings that achieve a reasonable blend of the two.

Without constraints, firms in concentrated free markets are able to exploit consumers. This may be in the form of straightforward profiteering or in more complex patterns of cross-subsidization to eliminate or block potential competitors. On the other hand, one does not have to be an economist to appreciate that regulation alone is generally not the best answer. The words themselves tell the story: to *compete* is active; it is positive; it is to do something. To *be regulated* is passive; it is negative; it is to be inhibited from doing something. Regulators do not tell executives how to run their businesses, and when businesses are run poorly, the regulators are often there to help bail them out at the customers' expense. Ex-regulators sometimes even become industry spokesmen, lobbyists, and special pleaders.

An interesting quirk in the debate over competition and regulation involves comparing the actuality of one process with the theoretical ideal of the other. Often those who attack the shortcomings of free markets contemplate an idealized picture of an efficient public agency promoting the public interest, while the critics of regulation compare it with an idealized version of perfect competition. In fact, both regulatory institutions and free markets are highly imperfect instruments of general economic welfare in the real world. Fortunately, both competition and regulation are opposite sides of the same coin. They are not mutually exclusive. They are both directed at the same economic objectives: efficient use of resources and the protection of consumers from exploitation. And in most real world circumstances, these economic objectives can be best pursued with a combination of regulatory and free-market tools, rather than by making an ideologically imposed choice between one approach or the other.

"Price cap" and "social contract" pricing proposals that afford integrated telephone companies pricing freedom for most of their service offerings in exchange for a promise to limit rate increases for essential basic exchange service to a predetermined amount (for example, indexed to the consumer price index [CPI]) are adverse to consumer interests in both short-term and long-term contexts. In the short term, provisions for programmed rate increases are almost certain to produce excessive rates for essential basic exchange services. In the long term, the inherent opportunities that such arrangements provide for cross-subsidy are almost certain to impair the evolution of potential competition.

Focusing first on the short term, these pricing approaches assume that existing basic exchange service rates have been set correctly in the first place. Virtually everyone agrees (albeit for widely differing reasons), that is not the case. Second, rather than rising with the CPI, basic exchange service rates should fall as the industry's technological and productivity gains outpace inflation, as the rate base depreciates, and as corporate tax rates and embedded interest costs decline. Third, in order to prevent anticompetitive cross-subsidies where common network facilities are used for both monopolized and competitive services, total system costs, as well as allocation mechanisms, must be subject to regulatory scrutiny.

From a long-term perspective, it must be recognized that competition is both a *process* and a *state*, and to complicate matters, a short-run adjustment in the process may impair the future state. In other words, competition is not merely a static condition; it has a time-shape. Pricing schemes in markets today will determine the competitive structure of markets in the future. If today's large and regionally dominant telephone utilities are granted a "social contract" that unleashes them to drive out or acquire smaller, less strategically advantaged, potential market rivals who have no local exchange monopoly to serve as a residual deep pocket, today's "competition" will lock up tomorrow's monopoly.

Pricing constraints and MFJ restrictions on the competitive business activities of the RBOCs in addition to protecting basic local service subscribers, will also enable independent competitors to gain a market foothold in some services. This has prompted the dominant exchange carriers to argue that regulatory restrictions that create an "artificial" opening for competitors are unfair and that deregulation should be pursued so that we can have a "level [competitive] playing field." In fact, however, deregulation in the name of "leveling the playing field" is more likely to level the ratepaying public and destroy the competitive playing field.

The notion of a level playing field has its intuitive appeal, especially in today's political arena. In considering telecommunications price deregulation, however, the level playing field analogy is misleading and wrong. Would the competition really have been more effective, more fair, and more efficient if David had not had the protection of a slingshot? If the Goliaths of this industry are given their level playing field, by eliminating or weakening regulatory restrictions against cross-subsidy, they will, of course, win all the "competitions"—no matter how efficient or inefficient they really are—because then the Davids won't even show up to compete, having encountered that most pervasive of all barriers to future competitive entry: remembrance of past disasters.

Regulatory controls and business restrictions for dominant local exchange monopolists are essential in order to prevent cross-subsidies and enhance competition in communications markets. Arguments that focus on leveling the competitive playing field miss the point. That is like arguing that David and Goliath should have arm wrestled first, with the slingshot going to the winner. Without line-of-business restrictions and effective exchange-service price regulation to prevent cross-subsidies, the potential for truly competitive market evolution will be destroyed.

» *Conclusion*

This discussion began as a commentary on new market pricing concepts, but turned ultimately to fundamental questions about competition and regulation. That was, perhaps, inevitable in that today's new pricing concepts purport to be more market-driven and less the product of detailed price regulation as we move toward competitive telecommunications markets.

Consideration of these interrelated topics has led to several conclusions. First, deregulation and market pricing freedom for exchange service carriers is not synonymous with competition. In fact, it is likely to impair the potential competitive process in telecommunications markets. Restraining new competitive entry has always, after all, been a major pricing objective of established dominant carriers.

Second, beyond their flashy wrappings, today's "new" pricing concepts are not fundamentally new. Value-of-service pricing was, for many years, the standard in telephone ratemaking. While considerable regulatory attention was given to cost-base pricing concepts in the late 1970s and early 1980s, that movement never matured to the point of general application in the telephone industry. For the most part, telephone service cost allocations and rate designs have remained arbitrary, with a large measure of market-oriented discretion by exchange carriers and with regulatory approval to recover all residual revenue requirements from noncompetitive basic exchange services. So-called social contract pricing would give a new formality to discretionary ratemaking for nonbasic services, while locking in current arbitrary rate levels plus programmed escalations for basic services—services that, in most cases, should now be entitled to rate reductions.

In short, as the telecommunications industry enters a new era, the pricing objectives and strategies of the major exchange service carriers, while laced with the appropriate rhetoric to accommodate the reigning political ideology of the day, are not fundamentally different from what they have been in the past.

12 » *A Regulatory Perspective*

Bruce J. Weston

To paraphrase Mark Twain, reports of regulation's demise are greatly exaggerated. The regulator's public policy oversight of most local exchange pricing is necessary now, and will be for the foreseeable future. This is because the local exchange provider is a monopolist, and monopoly consumers need regulatory protection from monopoly pricing, if not abuse. It's a strange commentary on the times that this point even has to be made, but there are those who contend that all or much of local exchange pricing should be constrained merely by currently or potentially operative market forces.

While it is difficult to fathom potential competition or contestable markets as substitutes for price regulation, the existence of real and effective competition would be another matter. But where is the competition? For the most part, it is anecdotal at best. After a thorough review of the matter, Judge Harold Greene recently found "that no substantial competition exists at the present time in the local exchange service." Judge Greene concluded that "only one-tenth of one percent of inter-LATA traffic volume, generated by one customer out of one million, is carried through non-Regional [Bell] Company facilities to reach an interexchange carrier." (*U.S. v. Western Electric,* 1987).

Of course, the average consumer knows that to be true without any assistance from the legions of experts who have addressed the subject. Consumers know that they basically have no local exchange alternatives other than the U.S. Postal Service and face-to-face conversation. The competitive candidates most often trotted out for some show of competition—cellular telephone, cable TV, and certain telephone equipment—are little threat to the local exchange company's monopoly.

Cellular telephone is nowhere near being an effective source of competition because of its high price and substantial dependence on the wire-line telephone companies for call completion, among other things. Ohio's recent telephone usage survey (Mount-Campbell, Neuhardt, and Lee 1987) showed that cellular had a very small percentage of the market; that small percentage, of course, does not mean that consumers abandoned their wire-line service for cellular. Indeed, cellular telephone may always exist merely as a supplement to regular, wire-line telephone service, but not as a replacement.

Cable TV is not telephone service now, nor will it be in the foreseeable future, if ever. (In fact, the more likely scenario may be that telephone companies will provide cable TV.)

As for telephone equipment, the foremost candidate is the PBX (private branch exchange). The PBX is a switch that an office can install to switch calls within the office instead of using the telephone company's central office switch. The PBX can switch calls within an office, but it must be connected to the outside world by the telephone company if the office's telephone service is to be of general use.

Peter Huber's 1987 work for the U.S. Department of Justice has become a popularly cited, albeit errant, support for the proposition that PBXs form an alternative telephone network. In reality, Dr. Huber's conclusion was merely that the interconnection of PBXs is "the major telecommunications challenge in the next decade," which is what he called "the geodesic network" in his report by that name (Huber 1987: 1.27).

Perhaps the liberties taken in citing Dr. Huber's report reflect that inapt title. For now and the foreseeable future, the findings in Dr. Huber's report would have been far better served by the title, "The Monolithic Network." The monolith is the local exchange industry, whose percentage market share was found by Dr. Huber to be well into the high 90s (Huber 1987: 1.20, 3.9). While the PBX may present competitive pricing questions for a specific exchange service (Centrex), it should not invoke questions about the exchange monopoly and the need for price regulation.

Local exchange service is a monopoly, and as such, its prices should be regulated, as opposed to deregulated. These prices include access service, directory assistance, operator services, custom calling, coin telephone, the rates that coin telephone resellers pay, Centrex, and exchange access for long-distance companies. On a service-by-service basis, some pricing flexibility may be warranted, such as for Centrex; but by and large, broad price regulation is still necessary.

The form of regulation most in the consumer's interest is rate base/rate-of-return regulation. There are variations that may be of interest, such as incentive regulation. But the great departure from rate-of-return models is price cap regulation, sometimes tagged with the elusive label, "social contract."

Price cap proponents criticize rate-of-return regulation for allegedly encouraging rate base inflation, or gold-plating, discouraging innovation, and being administratively burdensome. This could be called the Jekyll-Hyde effect. If the industry has undermined regulation, that apparently is due to its bad persona's uncontrollable impulses. In its good persona, the industry disowns responsibility for inflation of the rate base and for allegedly lackluster innovation. Under price caps, the good corporate persona will presumably prevail.

This outcome is just too improbable. The more likely result is that price caps will be less effective than rate-of-return regulation. In any event, the challenge of the future should be to improve rate-of-return regulation, not to reward its alleged failures.

What's wrong with price cap regulation? The list includes guaranteeing price increases according to an index (such as the consumer price index [CPI]) that has little to do with the cost of telephone service. Historically, in fact, increases in the CPI have exceeded increases in the cost of telephone service under regulation.

If a productivity adjustment is employed to deflate the CPI, minor progress is made; but such progress sacrifices the administrative efficiency that proponents of price caps commend. Indeed, the process of calculating the productivity deflator could look a lot like the cost of capital and cost apportionment determinations that were replaced.

Additionally, the cut-over price, or starting point for the price cap, could deny consumers the promised long-term benefits of, among other things, increased depreciation charges, which they have paid as part of their current rates. An example of this is a recent Central Telephone of Florida Five-year Plan (1986) wherein it was concluded, under the heading "Elimination of Rate Base Regulation," that:

> Technological changes and deregulation continue to decrease the Company's rate base investment per access line. With the Florida Public Service Commission's recent decisions to use one-time depreciation expense bookings to absorb excess company earnings, rate base regulation has become a concept that is no longer viable. The Plan, therefore, outlines possible alternatives to strict adherence to this concept of rate base regulation (Executive Summary).

The price cap also threatens quality of service. Dr. Douglas Jones, Director of the National Regulatory Research Institute, recently wrote of this as the " 'shrinking nickel candy bar' phenomenon" (Jones 1987: 20). In other words, the size or quality of the candy (that is, telephone service) may shrink, while the price is constrained.

There also is a legal question lurking here. If a commission were to order service quality improvements under the price cap, that order might be challenged by the carrier as unconstitutional on the basis that the price cap precludes a fair return on investment. This challenge could come after years of a carrier's enjoyment of profits (excessive or otherwise).

For most consumers, benefits of innovation are oversold. Mount-Campbell, Neuhardt, and Lee (1987) showed that most residential and business customers preferred POTS—Plain Old Telephone Service.

While someday consumers might want PANS (Pricey Advanced New Services), it is the availability of POTS at reasonable prices and adequate quality that most concerns customers.

As a final matter, a price cap regime might indulge two industry pricing objectives that are better left to full regulatory oversight. These are cost-shifting to reduce long-distance rates at the expense of local rates, and local measured service (LMS). With respect to the former, the British price cap system has allowed substantial local price increases to reduce long-distance rates. The general objective of such price restructuring is to eliminate the alleged "subsidy" of local service by long-distance service.

For too many, this subsidy proposition seems as genuine as Mom and Apple Pie. It is not, but a tremendous public relations campaign has been waged to sell the idea. First of all, the concept of "subsidized" local rates is a misnomer. The proponents of the subsidy argument would have it believed that a subsidy begins with the first cent that long-distance rates contribute to paying costs of the local telephone companies' networks. As stated in a National Regulatory Research Institute (NRRI) report, the substantial cost of a local telephone company's network (telephone wires, and so on) is "a common cost for the provision of local, toll, cellular and other services" (Racster, Wong, and Guldmann 1984: 151).

What this means is that long-distance companies need and use the local network to originate and terminate calls, just as consumers do in placing local calls. As network users, long-distance companies should pay part of network costs. Therefore, the process should be one of cost apportionment between the joint users of the network. However arbitrary that process might be, monopoly consumers are served where the regulator (and not the monopoly carrier) exercises the discretion to apportion the joint costs. In fact, the potential for arbitrariness is all the more reason for a regulatory balancing of the interests.

As for local measured service, a price cap could set the stage for premium pricing of flat-rate service as a means to migrate consumers to LMS. Consumers do not like LMS. That's partly why the following Bell System migration objective has not been achieved (AT&T 1979: Sect. B.2): "By 1985, nearly all business customers and a preponderance of residence customers will be charged for exchange service on a measured basis."

Apart from consumer animosity, important concerns have been raised that undercut the alleged cost justification of LMS. For example, it's not clear that local usage correlates with identifiable plant cost incurrence. Moreover, it is not clear that the cost of a given LMS call is sufficiently understandable and predictable to send accurate price signals to consumers. It's also not clear that LMS metering and billing costs are worth incurring for the undemonstrated benefits of LMS.

In conclusion, there is no crisis of regulation; the sky is not falling. Yes, Chicken Little is on the loose, but the chicken's facts are wrong. Regulators have much to be proud of, as does the industry under rate-of-return regulation. Deep breaths and clear vision are the order of the day.

» References

AT&T. 1979. *Interdepartmental Guidelines on Measured Service Implementation* (May).

Central Telephone Company of Florida Five-year Plan. 1986. Tallahassee, Florida.

Huber, Peter W. 1987. *The Geodesic Network: 1987 Report on Competition in the Telephone Industry* (Huber Report). Washington, D.C.: U.S. Department of Justice, Antitrust Division.

Jones, Douglas. 1987. *A Perspective on Social Contract and Telecommunications Regulation.* Columbus, Ohio: National Regulatory Research Institute.

Mount-Campbell, Clark; Jack Neuhardt; and Baumin Lee. 1987. *A Descriptive Study of Telephone Usage in Ohio.* Columbus, Ohio: National Regulatory Research Institute.

Racster, Jane; Michael Wong; and Jean-Michele Guldmann. 1984. *The Bypass Issue: An Emerging Form of Competition in the Telephone Industry.* Columbus, Ohio: National Regulatory Research Institute.

United States v. Western Electric, 673 F. Supp. 525, 540 (D.D.C. 1987).

13

» *Rate-of-Return Regulation in the Information Age*

Carl E. Hunt, Jr.

The primary reason that economic regulation is imposed on firms is monopoly power. Public utilities are firms that typically have a high degree of monopoly power and are heavily involved with the public interest. This combination of monopoly power and public interest, coupled with the potential inequities of price discrimination, has led to the unique American experiment of economic regulation of our public utility enterprises.

Figure 13–1 demonstrates the economic efficiency of monopolies and, consequently, one of the reasons for regulation. The shaded area shows the amount of excess profit (the profit that is above norm) earned by a monopolist. Since a monopolist, as with all firms, is constrained by the demand curve, the only way excess profits can be earned is by restricting output. For society, the result is higher prices at restricted quantities—for the monopolist, greater profit and economic power.

A goal of regulation should be to thwart the monopolist and force it to charge a lower price and produce a greater quantity.

FIGURE 13-1

» *Monopoly.*

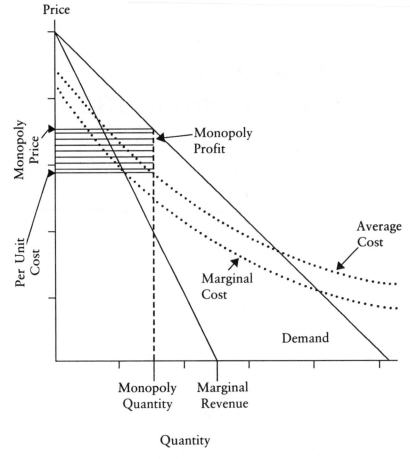

Figure 13–2 shows the difference between the monopoly price and quantity (*Pm Qm*) and the regulated price and quantity (*Pr Qr*). Ideally, the regulator will set the price equal to the point where average cost intersects the demand curve when both average cost and marginal cost are declining. Under these circumstances, the firm will earn a normal profit, since return on investment is included in the average cost curve.

Average cost for a regulated firm can be determined by the revenue requirement formula shown at the bottom of Figure 13–3.

FIGURE 13–2

» *Regulation of Monopoly.*

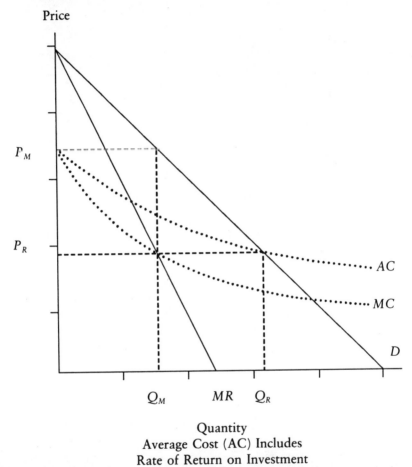

Quantity
Average Cost (AC) Includes
Rate of Return on Investment

Thus, in order to set a just and reasonable price, the information necessary to derive the revenue requirement formula is needed. Without this information, the regulator is reduced to arbitrarily choosing a price without knowing whether it tends toward the profit-maximizing monopoly price or toward a just and reasonable price, as defined by the intersection of the average cost curve with the demand curve. Therefore, rate base regulation, along with its complexities and data requirements, provides the regulator with the information needed to make reasoned decisions.

FIGURE 13-3

» ***Change in Average Cost for a Regulated Firm.***

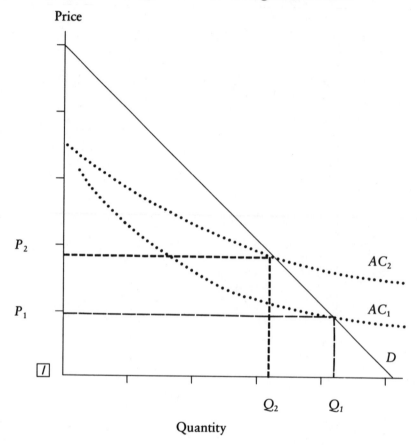

Revenue Requirement = Expenses + (Investment) Rate of Return

$$\text{Price} = \frac{\text{Revenue Requirement}}{\text{Quantity}}$$

$$\text{Average Cost} = \frac{\text{Total Cost}}{\text{Quantity}}$$

Some will argue that the marketplace should make those decisions, but the monopoly price is just as much a market-determined price as the competitive price. However, the difference is that firms in competitive markets are severely retained by the interplay of costs and demand. These firms are price-takers and cannot manipulate the price of their market.

On the other hand, firms with monopoly power are far less constrained. The monopolist is constrained by its demand curve, but has great latitude in determining its price, quantity produced, and level of profit.

A major criticism of rate base regulation is that it provides perverse economic incentives. The criticism is greatly exaggerated, as can be seen by examining the "Averch-Johnson effect" (Averch and Johnson 1962). In a nutshell, this theory states that because a regulated firm's profits are based upon the size of its rate base, it will have an incentive to increase the size of that rate base beyond the level that is economically efficient. A potential result of the Averch-Johnson effect is shown in Figure 13–3.

The firm increases the size of its rate base, shifting its average cost curve upward from AC_1 to AC_2. The regulator is forced to increase the price from P_1 to P_2, which consequently reduces the quantity produced from Q_1 to Q_2. Society suffers while the monopolist profits.

Each of the perverse incentives discussed by economists, whether it's the Averch-Johnson effect or some other negative incentive, results in driving the average cost curve upward and forcing the regulator to move price toward the profit-maximizing monopoly price. This is the case whether the disincentives pertain to expenses, investment, or rate of return.

An interesting note is that little empirical evidence has been found for the Averch-Johnson effect (Scherer 1970: 529–37; Arzac and Edwards 1979: 1053–69). It does not seem to be either prevalent or a major factor in regulation. This is the case for most of the so-called disincentives created by rate base regulation. Make no mistake, the incentives are real. They are theoretically supportable, and regulators should be aware of them. But as long as rate base regulation is practiced the way it has been historically, perverse incentives are not likely to be a problem. The reason is that incentives are not actions. A prerequisite is the ability or conduit to put incentives into action. The rate case process breaks the causation from incentive to action.

The rate case process is one of open information-gathering by experts. The utility's books and records are open to scrutiny. The process makes it difficult, but not impossible, for utilities to institute their perverse incentives. Informed commissions assisted by technical experts can determine what are allowable expenses and investments and the appropriate rate of return—thereby determining the location of the average cost curve, which is a necessary step to setting a just and reasonable price. Perversely, it is the absence of rate base regulation and the information it provides that will most likely increase the ability of utilities to behave in economically inefficient ways.

The cost of regulation is often cited as a major disadvantage of rate base regulation. Certainly, regulation is costly and time-consuming, but the costs are generally greatly overstated (Trebing 1987). This discussion will be brief, mentioning only a few items.

First, rate base regulation requires a considerable amount of information. Information is costly; however, the cost of gathering that information needs to be weighed against the alternative of not having it. Without the information, the regulatory authority does not know the location of the cost curves and cannot protect ratepayers against monopoly pricing.

Second, the cost of regulation usually can be passed on to ratepayers. The increased rate to Mountain Bell ratepayers in Colorado due to regulation is estimated to be 27 cents per month, or two-tenths of one percent of the ratepayer's monthly basic service rates (Colorado PUC 1986). This minimal cost needs to be weighted against the benefit of a doubling or tripling of basic exchange rates, which would occur without regulation. The cost benefit ratio is about 75. (A cost benefit ratio of 1 is usually sufficient to undertake a project.)

Third, many critics of rate base regulation claim that regulation is costly to the regulated companies and point to regulatory rate department budgets as an example. Again, these charges are overblown. Many of the costs are costs that would be borne whether the company was regulated or unregulated. All companies have rate departments. Large, complex companies such as telecommunications providers tend to have large rate departments. Eliminating regulation will not eliminate this function. A well-managed regulated company will combine its rate function with its regulatory function so that the additional cost imposed by regulation will be minimal.

An article by Kenneth D. Boyer (1987) supports the contention that the costs of regulation are grossly overestimated. Prior to deregulation, the cost of railroad regulation was estimated to be between $1.25 billion and $3 billion annually. Deregulation adherents suggested that after deregulation railroads would decrease price, owing to the elimination of burdensome regulation. This in turn would result in railroads gaining market share from other transportation modes. Boyer points out that postderegulation analysis invites vastly different conclusions. He states that the outside annual cost of railroad regulation was $93 million, or 0.32 percent of railroad annual revenues. In addition, railroad prices generally rose after deregulation, and hence railroads' share of the transportation market has decreased substantially since deregulation.

The latest fad in alternatives to the rate-of-return regulation bag is the Federal Communication Commission's (FCC's) price cap plan. Price caps impose a ceiling upon the rate a firm may charge. Price caps may be adjusted using a number of factors, such as productivity, inflation/deflation, or profits. The advantages of price caps are touted to be:

1. A carrier would retain profits garnered from efficiency increases, thereby encouraging greater innovation and efficiency.
2. Price caps decrease incentives to shift cost from more competitive to less competitive services.
3. They could reduce or eliminate incentives to inflate the rate base.
4. They may result in greater rate stability.
5. They are simpler to administer than the current cost-of-service approach.

An analysis of price caps needs to briefly return to the beginning of this paper, where the reasons for regulation were discussed. One aspect of today's telecommunications market needs to be kept in mind: if AT&T were not regulated, it holds such a dominant position that it would be a prime contender for an antitrust suit. The major concern in the long-distance market is whether AT&T's rivals will continue as minimal alternatives to AT&T. Prior to divestiture, the other common carriers (OCCs) received a substantial price break relative to AT&T. Since equal access, those price breaks are disappearing, and consequently, OCC profits are being squeezed. The danger to the FCC's competitive experiment is the demise of the OCCs. The FCC's answer to this problem is price caps, which reduces regulatory information and control and easily allows a movement toward monopoly pricing.

A stated purpose of price cap regulation is to relieve carriers from the burden of rate-of-return regulation, that is, to allow AT&T the flexibility to price more like an unfettered dominant firm. Allowing AT&T greater price flexibility provides an umbrella under which the OCCs can price, yet it also provides AT&T the pricing flexibility to control its rivals. The outgrowth is that profits for AT&T and the OCC's increase, thereby eliminating the potential failure of the FCC's competitive experiment in long-distance service.

The FCC's price cap proposal, which imposes a ceiling on average rates of capped services overall, abandons one of the primary principles of regulation—the prevention of undue price discrimination (FCC 1988). The super-index concept would allow AT&T to price using the inverse elasticity rule, or any other method, to price-discriminate among customers.

A major fallout of such a policy is that the customer access line charge, which has increased local rates, need not be applied to decrease message telecommunication service (MTS) rates. Thus, residential and small business customers would be forced to pay higher local service rates and long-distance service rates.

Some of the advantages of price cap regulation supposedly go to correcting weaknesses in rate-of-return regulation. As stated elsewhere in this paper, those weaknesses are generally overstated. The following are some comments about some of the weaknesses of rate-of-return regulation vis-á-vis price caps.

1. The telecommunications industry generally has been considered one of the most efficient, innovative industries in the country (Markey and Blau 1986). At a time when productivity in most U.S. industries was faltering, productivity in telecommunications was surging ahead. All this under traditional rate base regulation.

2. Because of the increased pricing flexibility, the possibility of less competitive services supporting more competitive services, far from decreasing, actually increases.

3. The incentive to increase the rate base is present with traditional rate-of-return regulation, as are a number of perverse incentives. The ability to act upon those incentives generally is not present in traditional regulation, owing to the openness of the process and the information requirements. The vast majority of studies conclude that utilities have little success instituting their perverse incentives.

4. The cost of traditional rate-of-return regulation is vastly overstated. It is not costless, but it provides information, a commodity necessary for intelligent action. It is a lack of information that will enable utilities to act upon their

perverse incentives and allow them the costly opportunity to price as monopolists.

5. A final claim about price caps is the administrative ease of their operation. To establish price caps will require traditional rate-of-return, cost-of-service analysis. To check the reasonableness of price caps at any point in the future will require traditional rate-of-return, cost-of-service analysis. The consequence is that price caps will not eliminate, but simply add to, traditional rate-of-return requirements.

Even without the need to periodically resort to traditional regulation, price caps do not reduce the complexity of ratemaking. They shift the type of regulatory analysis. Price caps could be described as a full employment act for economists. Appropriate formulas, estimating techniques, and data bases will need to be developed. Regulatory commissions can look forward to many hours of testimony about items such as the appropriate productivity measurement and price indices. This will, no doubt, increase the complexity of rate cases.

Regulators often are viewed as stodgy and opposed to changes to rate-of-return regulation. Numerous reasons are given for this perceived reluctance, including the economic, financial, political, and human capital investment that regulators have in rate-of-return regulation. This view does not comport with fact. The Civil Aeronautics Board was instrumental in deregulating the airline industry; the FCC has been instrumental in deregulating portions of the telecommunications industry and has pursued other regulatory reforms; the Federal Energy Regulatory Commission (FERC) has taken other substantive action to alter regulation of the natural gas industry by issuing Order 436, as well as other actions; the FERC is devoting additional resources to alternatives to rate-of-return regulation.

Just recently, the Kansas Corporation Commission established new rules to ease the regulatory burden on rural electric cooperatives. This action is similar to those of many other state commissions. Vermont, under the leadership of Vermont Public Service Commission Chair Louise McCarren, pioneered the concept of the social contract. When necessary, regulators have modified or moved away from traditional rate-of-return regulation. Additional examples include the use of future test years, make-whole rate cases, construction work in progress, incentive rates, settlement conferences, operating ratios, and indexing.

Regulators probably are no better or worse than any other group of people at responding to changes in their environment. Sometimes changes are anticipated and made, other times changes are resisted and forced by outside pressure. And sometimes change is appropriately resisted by regulators.

The fact that price caps do not appear to be an idea whose time has come does not mean that the FCC and other regulatory bodies should not explore alterations or alternatives to rate-of-return regulation. The divestiture changed AT&T from a capital-intensive utility to an expense-intensive utility. Consequently, AT&T's rate base substantially decreased. The result is fewer absolute dollars available for profits with a given allowed rate of return.

Based upon divestiture and regulatory changes over the past decade, a number of issues, concerning the type of regulation that should be asserted over AT&T must be addressed, including:

1. The market structure and degree of competitiveness in the interstate long-distance telecommunications markets
2. The potential for the current market structure to change to more or less competitive markets
3. The ability of AT&T to attract capital with its new position as an expense-intensive, as opposed to a capital-intensive, utility
4. The relationship of AT&T's rate of return to that of other companies with similar revenue-capital ratios
5. The relationship of AT&T's rate of return to that of other companies with similar revenue-expense ratios

If an analysis shows that AT&T requires nontraditional regulatory treatment, methods that allow for nontraditional regulatory control should be instituted. An important criterion for any new regulatory treatment is that regulators have sufficient information to make informed decisions. Lack of information or asymmetrical information places decisionmakers in the position of making poor decisions.

There are a number of nontraditional, less burdensome methods of regulation, such as operating ratios, that do not so dilute the information available to regulators as to make reasoned decisions impossible. Also, the FCC should pursue simplifying tariff filings and rate structure where possible. Price caps may be an appropriate tool, along with other tools, in a package that aims at rate structure simplification.

» *References*

Arzac, Enrique R., and Franklin R. Edwards. 1979. "Efficiency in Regulated and Unregulated Forms: An Iconoclastic View of the Averch-Johnson Thesis." In *Problems in Public Utility Economics and Regulation*, edited by M.A. Crew, pp. 42–54. Lexington, Mass.: Lexington Books.

Averch, Harvey, and Leland L. Johnson. 1962. "Behavior of the Firm Under Regulatory Constraint." *American Economic Review* 52 (December): 1052–1069.

Boyer, Kenneth D. 1987. "The Costs of Price Regulation." *Rand Journal of Economics* (Autumn): 408–416.

Colorado Public Utilities Commission. 1986. *Legislative Report.*

Federal Communications Commission. 1988. "Further Notice of Proposed Rulemaking in the Matter of Policy and Rates Concerning Rates for Dominant Carriers." CC Docket No. 87–313, FCC 88–172 (adopted 23 May 1988).

Markey, David J., and Robert T. Blau. 1986. "Is the AT&T Consent Decree Strangling American R&D?" *Telematics* 3, no. 8 (August): 3–8.

Scherer, Frederic M. 1970. *Industrial Market Structure and Economic Performance*. Chicago: Rand-McNally and Company.

Trebing, Harry M. 1987. "Regulation of Industry: An Institutional Approach." *Journal of Economic Issues* 21, no. 4 (December): 1707–1737.

» PART IV

Perspective of the Analysts: The Price Drivers

14 » *Deregulating the Local Exchange*

John T. Wenders

Too little attention has been paid to the whole issue of local competition. Aside from the fact that the industry has been preoccupied with divestiture and access charges, at least four reasons lie behind this failure. First, in many cases, much local competition has been simply made illegal by the granting of exclusive franchises. Second, the underpricing of local service, especially local residence service, has thwarted competition from unsubsidized potential competitors. Third, what local competition did appear was called "terminal equipment competition" without any appreciation that it was not "terminal" and had important implications for the entire way in which the network is provisioned. Finally, there has been a tendency to think of local competition largely in terms of bypass competition.

In order to understand the real potential for local competition, one must look at the broad impact of the role of technological change in the provision of local telephone service. Specifically, technological change has altered the way in which local service is provided by making it economical to substitute cheaper multiplexing, switching, and trunking for more costly access lines. This development has lowered the minimum efficient size necessary for the provision of local service and destroyed the natural monopoly characteristics of local service—if there ever were any.

The thrust of this paper is that the vulnerability of the local telephone market to competitive entry—entry made possible by improved local multiplexing, switching, and trunking technology—makes it possible to rely on competition rather than regulation as the major disciplinary force in local telephone markets. Further, the presence of such local competition makes flat-rate pricing unsustainable, thus making any squabbles about the neoclassical benefits and costs of measured service irrelevant.

The speed with which the discipline of local competition can be substituted for regulation depends on the speed with which local "costs" are unloaded from toll prices—thus making local service stand on its own—the speed with which the local companies' capital recovery problems are dealt with—thus giving local companies an incentive to trade regulatory protection for competition—and the speed with which state regulators eliminate the legal barriers to local competition.

Yet there is much that can be done immediately. There is no reason why the local network should not be opened up to free and fair competitive entry. State toll and Centrex are already effectively competitive—the former by the presence of actual and potential interexchange carriers (ICs), and the latter by private branch exchanges (PBXs). If single and multiline business and PBX trunk services are priced on a nondiscriminatory basis, the reselling of Centrex will provide competitive discipline to these services, too.

Since there is much confusion and ignorance—especially at the U.S. District Court in Washington—about the concept of so-called natural monopoly and the meaning of competition, this paper begins with a simple exposition of both the positive and normative aspects of these concepts. It then turns to a brief history of local competition, followed by an exposition of the way in which it is presently proceeding and, finally, reasons why the current state of local competition makes flat-rate pricing unsustainable.

» *Natural Monopoly*

The conventional wisdom is that local telephone service is regulated because it is a natural monopoly. My view is that local service ended up being a regulated monopoly not because it was "natural," but because of some unique events and circumstances in the early history of the industry—circumstances that have little relevance to the present state of the industry—and because regulation subsequently sustained and solidified these early events. There is no evidence, other than nirvanic musings about economies of scale, that there ever was anything "naturally" monopolistic about local service. As John Haring (1984: 22–23) has succinctly stated in another context:

> [The natural monopoly argument for regulation] amounts to an argument that if we knew we had a natural monopoly (which we do not know) and if a natural monopoly had adequate incentives to innovate (which it may not, given effective profit regulation) and if a regulated monopoly could be relied upon to price efficiently (which historically it did not), then regulated (and in some cases closed) monopoly would be more efficient than alternative modes of organization.

By common usage, the natural monopoly argument for public utility regulation is that because economies of scale and scope exist, regulation will improve economic welfare. This argument presumes that (1) there are such things as objective economies of scale and scope and that they are discoverable independent of the market organization that produces them, and (2) the existence of these economies necessarily means that regulation will improve economic welfare. It is useful to deal with each of these assumptions in turn.

Economies of Scale and Scope

Natural monopoly arguments implicitly presume that "true" costs, including objective economies of scale and scope, are independent of the market organization from which they emerge. This is questionable. Empirically, objective costs result from choices made by producers. Such choices are necessarily made in the context of the incentives provided by the existing institutional setting, so that the character of the market and its institutions will influence what kind of cost data we observe flowing from the choice process.

Thus, since objective costs can only emerge from the choices made in a market context, it is not correct to presume that the costs that have emerged under regulation are the same as those that would have emerged under competition. This would only be possible if we were able to set up these two forms of market organization at the

the same time, under the same conditions, and then observe what kind of costs emerged under each. The natural monopoly argument presumes that the costs that have emerged under regulated monopoly are the same as the ones that would have emerged in an unregulated market.

Given that we cannot run two parallel experiments to discover the alternative characters of costs that would emerge under competition and regulation, what does one know about these two alternative forms of market organization that might give us some hints as to what might emerge? The hypothesis of this paper is that the costs that would emerge under regulation would be both higher and more likely to display economies of scale than those that would emerge under competition. There are three reasons for this. First, since both public utility regulation in general and telephone regulation in particular have had strong "cost-plus" elements, the incentive to minimize costs is considerably weakened under this form of market organization. Thus, from an incentive standpoint, one would expect costs under regulation to be higher. Second, the higher costs of operation under regulation have been confirmed empirically. There is now a good deal of evidence from a variety of industries that once protection from competition is removed, formerly regulated firms almost always go through a period of painful cost reductions, sometimes resulting in bankruptcy. Finally, with all output being supplied by a protected monopolist, the technology chosen and technological change itself are both likely to be directed toward the single-firm monopolist's situation, resulting in production methods that display appropriate economies of scale. Technology and its rate of change are not exogenous. Conversely, I would expect that a flatter average cost curve would emerge under a competitive market organization, merely because smaller firms would have a greater demand for this type of technology. In short, one might find that natural monopoly cost conditions are not a reason for regulation, but result *because* the market was regulated as a monopoly.

Connecting Costs and Prices

The natural monopoly argument for the regulation of markets is almost always stated like the following: "When economies of scale and scope are present, a single-firm regulated monopoly *can* produce in a more efficient way than several competitive firms." Alternatively, the argument states that in the presence of the economies of scale and scope, a regulated monopoly will avoid the wasteful duplication of facilities. These are non sequiturs. They are purely supply-side arguments, and it is well known that efficient production requires *both* that production be carried out in the most efficient way and that price be set correctly. Yet, cost arguments say nothing about how price will be set in a regulated monopoly environment. Further, the argument almost always says that economic efficiency *can* be improved by regulation, but almost never goes on to show how regulation *will* be more efficient by ensuring that either costs will be minimized or the price will subsequently be set correctly. Even if competition produces, from a theoretical standpoint, an apparently wasteful duplication of plant, this apparent inefficiency is likely to be outweighed by the superior price and cost-minimizing discipline offered by an unregulated market.

This is natural monopoly argument's fatal flaw: it presumes that the political process that governs the behavior of regulators will give *both* the regulators and the regulated the incentive to produce *and* price in the theoretically most efficient way. This is a classic, and commonplace, problem of using the unattainable as a guide to economic policy. At a general level, it is folly to use idealized states of the world as guides to economic policy in the absence of a real world mechanism by which this idealized state can be reached.

The decision to regulate a market cannot be made by looking at idealized views of the world. One should not compare the state of a market with an idealized market and then conclude that, since the idealized market has not been achieved, the market therefore should be regulated. This assumes that idealized regulation approximates idealized competition. Neither idealized competition nor idealized regulation are before us. The real questions are how both competition and regulation are likely to unfold, and whether or not leaving the market alone will result in performance superior to what is likely to result from regulation.

It might be noted in passing that arguments about the nonsustainability of natural monopoly, with the implication that protected monopoly would be better, are also beside the point because they presume the alternative is a perfectly regulated natural monopoly.

» *Competition*

How one views the status of local competition is determined by how one views the role of competition in our economy in general.

If one believes only in perfect competition, with all other forms the proper target of regulation and antitrust, then, indeed, lawyers, judges, and regulators will be busy. If, however, one views competition as a dynamic, disequilibrium process by which resources are attracted or repelled by prospective profits, such as presently takes place in most of the unregulated economy, then one will conclude that competition can provide a great deal of discipline in local telephone markets, allowing much of this market to be deregulated.

The problem is that many economists, most of them laymen, and all lawyers (and, apparently, judges) have such an idealized view of the competitive process that real markets seldom measure up to it. Their view of the competitive process is idealized in at least two related ways. First, they presume that the competitive process should take place much faster than it actually does, and second, they tend to view competition in long-run equilibrium terms.

At the risk of oversimplying, it is useful to briefly describe how competition works, from both positive and normative viewpoints.

A competitive market works in the following general way. When prices are above the long-run supply price (what this paper sometimes calls *market* long-run marginal cost to distinguish it from the *firm's* long-run marginal cost), abnormally high rewards will be earned by new resources, and thus resources will be attracted to the market in question. In addition, the lure of potentially high rewards will induce suppliers to try new products and means of production. Thus, the *possibility* of above-cost pricing provides a lure for innovation, and at the same time the fact of above-cost pricing encourages competitive entry, which will ultimately prevent above-cost pricing from lasting.

Similarly, when price is below the long-run supply price, the market will not be able to attract or hold productive resources. As capacity wears out and other resources become aware of better alternatives in other markets, resources will leave the market and will not be replaced. This decline in productive capacity will ultimately force prices to rise toward long-run marginal cost.[1]

This Marshallian process is described by all good introductory texts, but the theory tends to submerge a good bit of useful economics. As stated, this process is timeless. How long it takes for the entry and exit of capacity to take place is simply not specified by the theory; nor can it be, and this fact admits both error and mischief.

Most economics texts implicitly suggest that the process *should* take place rather rapidly in calendar time. If it does not, this is taken as prima facie evidence that the market is not "competitive" and therefore is fair game for "corrective" government action in one way or another. As with natural monopoly arguments for regulation, the tactic is to compare "imperfect" markets with "perfect" government behavior, and then jump to the non sequitur that because the latter is superior, *actual* government intervention will improve matters.

It is more realistic to view the competitive process as grinding rather slowly in the real world. Of course, it is "slow" relative to the instantaneousness of theoretical competition. But the facts of economic life are that resources are specialized and durable in the short run and therefore entry and exit cannot be instantaneous in any real world situation. Thus, in trying to apply the theory to the real world one must deal with the fact that markets will usually be in long-run *disequilibrium* in terms of the model being applied.

In the process of allocating resources by encouraging entry and exit, competition is said to promote economic efficiency. In simplest terms, maximum economic efficiency results when voluntary exchanges, each of which leave both parties better off, are maximized in any market. It is well known that maximum economic efficiency results when prices are cost-based, and this is why many economists, including this one, have recommended such pricing in regulatory proceedings. Competition, therefore, produces economic efficiency by promoting cost-based pricing.

By cost-based pricing, economists usually mean long-run marginal-cost pricing. In the short run, because markets are almost always in competitive disequilibrium, prices may be either higher or lower than long-run marginal cost. The only sensible way to reconcile this disequilibrium with economic efficiency is to view it as a temporary and *necessary* feature of the competitive process. It is temporary because the positive or negative rewards associated with such disequilibrium will necessarily induce corrective resource flows. It is necessary because it is the pursuit of better rewards, either within or without the market in question, that causes such resource flows in the first place.

Of course, competition not only promotes cost-based pricing but will promote *minimum* cost-based pricing. Prices will tend to move toward the minimum cost of production *at the margin,* and any firm that does not produce in the most efficient way will be elimininated by the competitive process in the long run. This, of course, lies behind the often stated rule that the role of public policy should be the promotion of competition, not the protection of competitors.

However, the idea of competition as a promoter of economically efficient cost-based pricing really does not correctly portray the entire role of competition in the marketplace. As noted earlier, the kind of cost-based pricing most often recommended by economists in regulatory proceedings is *long-run* marginal-cost pricing. This kind of cost-based pricing is equilibrium pricing in the sense that it represents the central tendency toward which prices will gravitate in a competitive market. There is nothing wrong with this idea as a long-run equilibrium concept that is useful in reasoning about market behavior. This idea of pricing, however, loses sight of both the dynamic forces that propel the market in the right direction when prices do indeed depart from costs, and the incentive that the possibility of above-cost pricing provides for entrepreneurial activity. When a market is looked at solely from the equilibrium perspective of long-run cost-based pricing, all departures from such pricing are bad and invite intervention to try to straighten matters out from a long-run perspective. This view fails to recognize that such departures are a normal and vital part of the functioning of a competitive market. Indeed, the departure from long-run cost-based pricing provides the incentive for competitive forces to operate, forces that both induce innovation and move the market toward cost-based pricing.

Thus, in a sense (and this is where the goal of long-run marginal-cost pricing can be misleading), the dynamics of the marketplace are brought into being by non–cost-based pricing. In the long-run equilibrium view of the market, non–cost-based pricing is bad. But this is true only from the long-run perspective. In the short run, it is this very non–cost-based pricing that provides the dynamics of a self-correcting marketplace. From the long-run perspective, every economic distortion brought about by non–cost-based pricing contains the seed of its own destruction—a seed driven by the competitive profit motive.

In addition, looking at any market from the standpoint of long-run cost-based pricing gives a distorted view of what a market will look like in disequilibrium. It obscures the fact that any competitive market, at any point in time, will contain a diversity of firms. Some will be doing a good job of providing what the market wants at a level of costs that leaves them with high "profits." Others will be doing less well, and in order to survive, these will have an incentive to imitate, and compete successfully against, those that are doing better. At any time each firm may have some advantage over others, and this is one of the desirable features of competition. This diversity of talent and luck is overlooked by the long-run, price-should-equal-cost equilibrium view of the market, even though such a long-run view may perfectly adequately describe the direction in which the market is going. But this long-run equilibrium view of the firm cannot be used to judge the short-run performance of a market that, by definition, is almost always in long-run disequilibrium.

Thus, in any competitive market we should expect to find the following. At any point, we should find firms of differing strength and advantage, depending on their respective amounts of luck and talent. The center of gravity for prices will be current long-run marginal costs, and, of course, current long-run marginal costs may change over time and in fact may never be reached in a dynamic market such as the telecommunications industry, where demand and technology are constantly changing. But when prices depart from long-run incremental costs, competitive forces will be set in motion that take prices in the direction of such costs. The potential for above-cost pricing will give an inducement for beneficial entrepreneurial activity in the marketplace. In this way, competition both regulates and stimulates any market.

Perhaps the major point to be made is that competition is a market process that reveals in a general way what *market* long-run marginal costs are. Indeed, from this viewpoint, it is probably impossible to discover such costs *ex ante*; they may be only a theoretical reasoning point to explain what one observes and therefore may not be embedded in any data. Such costs are probably not something that can be ascertained by looking at the firms in the industry at any point in time.[2] Such costs are the result of the competitive process, not an input into it. In a somewhat perverse way, it is useful to define market long-run marginal costs as the level to which the competitive process will drive price in the long run.

Before leaving this discussion of the disequilibrium nature of the competitive process, it is important to point out another feature of this process that is relevant to the current deregulation efforts in the telecommunications industry. The resources and institutions necessary in a truly unregulated market are quite different from those relevant to a regulated market, *even if price were equal to long-run marginal cost in each.* A competitive market will require firms to be prepared for the problems and opportunities that will be encountered in disequilibrium. This requires different human and physical capital and institutions than would be required of the firm in a regulated environment, even if such regulation was "perfect," that is, price being equal to long-run marginal cost was the outcome.

The other side of this coin is that regulation will necessarily have stifled the development of this kind of disequilibrium capital and institutions. Thus, not only has regulation necessarily suppressed the actual alternatives that are ordinarily present in a competitive market, it will also have suppressed the mechanisms by which these alternatives develop, appear, and disappear. But this does not mean that such alternatives will not appear with deregulation, once the disequilibrium capital and institutions have been given a chance to do their job. This is one of the great hurdles that must be overcome to effectively deregulate an industry: at a time when regulation has not only suppressed actual competition but also suppressed the development of the unique capital and institutions that make the competitive market work in the inevitable state of disequilibrium that will exist most of the time, regulators and the public must somehow be convinced that competition will be effective.

Finally, when a market is deregulated, even if long-run marginal-cost pricing has been the basis for past regulation, it invariably will be left in a state of competitive disequilibrium. Some temporary welfare distortions may appear, but they should not be used as a reason or excuse for continuing regulation; this short-run situation will be dealt with by the competitive forces that will appear.

The Role of Entry

As the above discussion indicates, the key to the control of this distortion of market power, in any market, is the free entry of capacity by completely new firms, firms established elsewhere but new to the market in question, or firms already established in the market.

It is important to realize that the condition of entry, like competition, is a *market* phenomenon; it is not dependent on the assessment of the strengths or weaknesses of any individual participants in that market. Easy entry ensures that a market will be competitive, but says nothing about which competitors will be the vehicles for that competition. Conversely, the role of public policy should be the promotion of competitive entry, not the protection of competitors, actual or potential.

Actual entry provides discipline by adding to industry capacity whenever price is above incremental cost. The existence of price above cost means that abnormally high economic profits are being earned. This makes a market vulnerable to entry, which will add to market supply, eventually drive price to the vicinity of cost, and improve the economic performance of the market in question.

Potential entry provides discipline to existing suppliers by making them take entry into consideration in deciding how high to price above cost. Potential entry may provide more discipline in a market than actual entry because it is not dependent on the presence of other competitors. Unless an existing firm that has 100 percent of the market also has some cost advantage over actual or potential entrants, it will not be able to raise price much above cost without attracting a host of entrants whose supply will drive price back toward cost. As a matter of fact, a good case can be made that the existence of a 100 percent market share may be an indication that the market is competitive, not monopolized, as some economists and all lawyers seem to think. Even a firm with 100 percent of the market is subject to the competitive discipline of potential entry, and the fact that no entrants have appeared indicates that the firm is behaving competitively, not monopolistically.

Schumpeter some time ago (1950: 85) put it in the following way, albeit in a slightly different context:

> It is hardly necessary to point out that competition of the kind we now have in mind acts not only when in being but also when it is merely an ever present threat. It disciplines before it attacks. The businessman feels himself to be in a competitive situation even if he is alone in his field.

More recently, Baumol, Panzar, and Willig (1982: 222) have been more explicit:

> [F]reedom of entry, indeed the mere threat of incursions by entrants into the market may effectively discipline the monopolist, even if entry is never successful. It can force the monopolist to curb his avarice and forgo profits he might otherwise have enjoyed. . . . Potential competition can also force the monopolist to produce with maximal efficiency, and to hunt down and utilize fully every opportunity for innovation. Perhaps most surprising of all, it can induce the monopolist to institute those . . . prices which welfare theory has shown to be requisites of Pareto optimality under a profit constraint. In short, the threat of entry can force virtuous behavior upon the monopolist, for if he behaves badly his monopoly becomes vulnerable. In our analysis, it is freedom of entry alone that is capable of accomplishing all these things.

There is ample evidence of the persuasive effect of potential entry in all aspects of the telecommunications industry, including the local market. Local telephone companies began to recommend in the early 1980s that state toll prices be reduced. These recommendations were made not because actual competitors were making huge in-roads into the intrastate toll business, or because any competitors had appeared, but because local telephone companies began to realize that their toll prices were way above cost and that the FCC's pro-competition stance was removing many of the regulatory barriers to entry into the various toll markets. Similarly, AT&T proposed interstate toll reductions before actual competitors appeared on most routes. Thus, the discipline of competition on sellers was felt long before actual competitors appeared. And as this paper shall point out in more detail below, there is strong historical evidence that local telephone prices will be disciplined by potential entry as well.

» *The Early History of Local Competition*

When one looks at the way in which the U.S. telecommunications industry has evolved, it is quite clear that the structure of the industry did not emerge from any purely economic competitive process. The genius of Theodore Vail of AT&T was that he was able to perceive what the political marketplace required and was able to devise an accommodation that was beneficial to both the body politic and AT&T. This was important both during the period of unregulated Bell monopoly (1876–1894) and in the later years when Bell successfully traded an acceptance of regulation for antitrust immunity, which allowed it to merge with its independent competitors as a way of reestablishing its dominance. It seems clear that if Vail had persisted in his efforts to keep AT&T fully integrated, and unregulated, the political mechanism would have produced either antitrust action or nationalization, or both.

What follows is a credible interpretation of events during the first half-century of the industry (Brock 1981: chs. 4 and 6; Bornholz and Evans 1983: 7–40). During the period of unregulated monopoly—a monopoly created and held primarily by patent protection—the Bell System gained an important head start on its competitors. Service was initially offered only in the central cities, and Bell was able to gain, and subsequently keep, exclusive franchises there. City governments became an important regulatory barrier to entry by later entrants.

Once Bell's basic patents expired, independent competitors rapidly entered, grew, and thereby performed a classic welfare-improving function. The independents' share of the market went from

6 percent in 1894 to 44 percent in 1902, to 51 percent in 1907. Further, the independents made substantial gains in the toll markets, carrying some 20 percent of the traffic by 1907. Bell's subscriber prices fell by at least one-half from 1894 to 1909, and this fall in price was about the same both where Bell faced local competition and where it did not. Bell's return on investment averaged 46 percent from 1876 to 1894, but fell to 8 percent from 1900 to 1906. By the turn of the century, 45 percent of cities with more than 4,000 telephone subscribers had two or more telephone companies supplying local service (Brock 1981: chs. 4 and 6; Bornholz and Evans 1983: 7–40).

It is significant that most of the growth of the independents came not by actually taking subscribers from Bell, but by offering service to cities, parts of cities, and rural areas (primarily in the Midwest) that were untouched by the previous Bell monopoly. Yet, this growth of the independents was a significant competitive threat that caused Bell's prices to fall in both competitive and non-competitive exchanges. *Both* actual and potential competition worked. One cannot think of a better example of how actual and potential entry can discipline monopoly behavior.

Also during this period, there is the first evidence of Bell successfully playing the political game of reducing local rates in return for the maintenance of exclusive franchises. The formation and maintenance of such exclusive franchises was particularly important to Bell in keeping monopoly control of several large cities, such as New York, Boston, and Chicago. There is no question that the early urban franchises acquired by Bell proved to be significant deterrents to later competition.

In the early years of development—up to about the early 1900s—a large subscriber base was apparently not necessary for an economically viable telephone exchange. This seemed to be due, at least in part, to relatively small communities of interest by subscribers, as evidenced by the fact that there was local telephone competition for only 8–13 percent of subscribers. It was primarily businesses that had duplicate service during this period. However, it is also clear that increased population mobility, urbanization, and suburbanization—all attributable to both rising incomes and improved methods of transportation—increased both business and residence subscribers' communities of interest and made wider calling more valuable. These developments, in turn, made the availability of good toll calling and interconnection among the various exchanges more valuable. Thus, while they may not have been necessary from the standpoint of getting started, it was probably true that good toll calling and interconnection had to be offered in order for a local exchange to remain competitive after 1907.

The independents, as well as Bell, took a strong stand against interconnection. This was probably suicidal on the part of the independents. In their heyday, the independents may have had the political clout to force Bell to interconnect, but because of their rivalry with the hated Bell System they chose not to use it. Had they done so, the merger wave initiated by Vail might have been less viable, and the independents might have become equal competitors to Bell.

On the toll front, it was, again, key patents—the Pupin coil and long-distance repeaters based on the vacuum tube—that gave Bell a strong advantage. The promise of good, clear, and long-distance toll connections, made possible by these patents, gave Bell a further competitive advantage in the local service market vis-à-vis the independents. Further, the independents' efforts to establish competitive toll networks were greatly hampered by their inability to gain access to many of the major cities because of exclusive local Bell franchises. Again, these circumstances greatly aided Vail's plans for both internal and merger growth.

In the early years of the industry, switching and trunking—the so-called traffic-sensitive parts of the network—were, in retrospect, relatively expensive. The legendary telephone operator at a cord board performed all switching. The use of switching machines of any kind was not widespread until the 1920s. Further, the multiplexing of trunks did not become widespread until after the Second World War.

The relatively high cost of switching and trunking, combined with policies against interconnection, made it necessary for potential local entrants to bite off relatively large populations and geographical areas—encompassing viable communities of interest— in order to be successful. How important this was for the development of the industry is uncertain. However, as seen below, when interconnection is possible the great reduction in the cost of switching and trunking reduces the minimum efficient size of providing local service and makes the local telephone market extremely vulnerable to competitive entry.

The presence of predominantly flat-rate local service, made necessary by the cost of measurement, undoubtedly made interconnection more difficult to negotiate. The effect of this on the development of the industry is, again, uncertain. But the present possibility of easy and cheap measurement undoubtedly makes interconnection easier.

These events and circumstances suggest that the following would have preserved a competitive local telephone industry and might even have forestalled the detailed regulation of the industry that Vail traded for continued Bell dominance:

1. The absence of key patent monopolies by Bell
2. Mandatory interconnection
3. No exclusive franchises
4. A separation of toll and local services
5. Prevention of widespread merger activity

These observations, of course, have the benefit of 20/20 hindsight. But these lessons of history bless us with strong evidence that local service can be offered in a competitive and largely unregulated format if these conditions prevail, and as will be argued in some detail below, these preconditions for competition and deregulation of local service are certainly presently within reach.[3]

» *The Mechanics of Modern Local Competition*

With the development of electronic switching, the multiplexing of channels, and microwave, satellite, and fiber-optic transmission, the cost of switching and trunking has come down greatly, relative to the cost of access lines. Thus, it has become more economic to provision local service using relatively shorter access lines and more subscriber line carrier switching and trunking. This is the fundamental point made by Peter Huber (1987: ch. 1) in his recent report to the Justice Department (DOJ) on competition in the telecommunications industry. Just as technological change has brought the computer closer to the user, it has brought switching and multiplexing closer to the telephone end-user. In essence, a new class of central offices, really Class 6 offices, has appeared, largely in the form of PBXs. There are already more business lines connected to PBXs than to Class 5 end-offices. The competitive significance of these developments is that the minimum efficient size of competitive entry has been significantly reduced over the past few decades.

The Role of Competition

·We are now in position to see how competition can attack local companies, to the extent that they are not protected by regulation. We shall also see that, even with regulatory protection, local companies are *already* being so attacked.

The first casualty of the loss of regulatory protection would be statewide average pricing. Even with the toll-to-local subsidy, some subscribers are overcharged, owing to averaged, flat-rate pricing; under free entry, there would be an incentive for competitors to try to pick these subscribers off, even if there is no technological change in the way local service is provided. Under the monopoly provision of local service, averaging has taken place in both the access and usage dimensions. Thus, those most likely to be overcharged by this averaging process are the short-loop, low-usage subscribers. Because of the practice of charging residence subscribers less than business subscribers, the latter probably would be more vulnerable to competitive entry.

But the emergence of such competition is much more fundamental; it is not simply a reaction to averaged pricing. Even though local service is subsidized relative to historic costs, technological change has made these historic costs higher than current costs. As we have seen, this is because dramatic shifts in multiplexing, switching, and trunking technology have lowered the cost of providing local service and reduced the value of a good deal of local plant. There was an alteration in the way in which local services are provided: using the relatively expensive service, access lines, more economically, and expanding the use of the relatively cheap service—multiplexing, switching, and trunking. This is access-line–conserving because access lines have become relatively more expensive, and thus, local service should be provided with shorter access lines. This was the way in which competition attacked the local companies, and it is certainly the way in which competition is presently attacking these companies—witness the accelerating battle over the provision of local services by landlords and developers using their own inside wire and PBX-type switches.[4] This is exactly the kind of competition that is now being fought out between PBXs and Centrex.

The emergence of access-line–conserving competition is an indicator of the kind of competition that would emerge in the larger local telephone network if it were permitted. Competition has come to the fore first in the provision of PBXs, local area networks (LANs), shared tenant services (STS), and inside wire. This is because relatively little of the toll-to-local subsidy was directed at the business sector, where this competition is the most prominent, and because common ownership made it cheap for the competition to organize, especially *ex ante*.

It is more difficult for local competitors to bite off pieces of the local network not under common ownership, although Centrex resellers are now doing this profitably. However, with the regulatory barriers to such competition lowered and a drying-up of the present toll-to-local subsidy, one would expect this form of competition to become much more prevalent. Telecommunications entrepreneurs would bite off a portion of the local network, install a remote multiplexing and switching device like a PBX, and connect to the present network via PBX-like trunks. Again, this says that the emergence of PBX, LAN, and STS competition is much more than simply a response to the statewide averaging of local prices. It goes to the heart of the way local service is being provided.

These possibilities also point out the problems associated with thinking of competition as simply alternative ways of providing like services. One of the checks on the overpricing of access by local companies is the provision of multiplexing, switching, and trunking away from present central offices, as a substitute for access. Given that toll services provided through the typical local company are also priced above cost, such arrangements also have the ability to concentrate toll calling to take advantage of bypass agreements with toll carriers. Contrary to the way in which many local companies and regulators think, PBXs are not simply a piece of terminal equipment. They are not terminal. They are another switch in the switching hierarchy of the network.

Given the rapid pace of technological change, and given a drying-up of the toll-to-local subsidy, this analysis suggests that the local companies can be completely deregulated if they are *unprotected* and thereby subject to the discipline of actual and potential entry. This paper next discusses some of the conditions under which such deregulation would be most effective.

Implications for Local Service Pricing

It is useful to point out the way in which such competition would alter local service pricing. Local companies will respond to competition by pricing their services *as if* they were provided in this way and by installing similar remote switching devices or concentrators when it is economical to do so. This, of course, is nothing more than a restatement of the age-old principle that, under competition, prices will be determined by the cost of production *at the margin,* even if the provider has costs that are higher or lower than marginal costs.

Further, as noted briefly above, when economic and regulatory barriers to entry were higher under old methods of local service provision, it was possible to engage in a great deal of averaging of prices. Competition will also put an end to this practice and force much more unbundling of local service prices. The pressure for measured service is one kind of deaveraging that is already being felt.

In the extreme, the new methods of provisioning local service might mean that relatively few loops will be connected directly to switches in present central offices. The employment of attendant concentrators and switches remote from present central offices means that present switching machines will begin to take on the characteristics of local tandem offices, which are entirely traffic-sensitive. The upshot will be that local measured service (LMS) will be an even more desirable way of pricing the network of the future, and its implementation will be hastened by the spread of local competition. Conversely, the spread of measured service will aid the spread of competition and the beneficial discipline that competition will bring to the local telephone market.

» The Death of Flat-Rate Pricing

With a few notable exceptions, such as the cities of New York and Chicago, local telephone service in the United States has been largely provided under flat-rate pricing. In the late 1970s, AT&T laid out a plan to convert most of the Bell System's flat-rate pricing to local measured service by the mid-1980s.

Economists quickly pointed out that LMS could be analyzed with simple neoclassical benefit/cost tools (Mathewson and Quirin 1972: 335–39; Mitchell 1978).[5] Under flat-rate pricing, efficiency in the usage market is distorted by proving usage at a zero price when it is not costless to provide. This produces a classic welfare-loss triangle that can be eliminated only by measuring and pricing usage. This, in turn, becomes the benefit available to LMS. On the other hand, flat-rate pricing does not require that resources be devoted to the measuring and billing of local usage, and this is the cost of LMS. The simplest and narrowest benefit/cost test for LMS is to weigh the former against the latter.[6]

The results of such benefit/cost tests have been mixed. Bridger Mitchell (1978) and James Alleman (1977) found LMS to be of net benefit. Recently, James M. Griffin and Thomas H. Mayor (1987), using a much broader and more complete test, also found LMS to be preferable to flat-rate pricing, on efficiency grounds. On the other hand, in a widely publicized study, Park and Mitchell (1987) provide a very interesting discussion of the peak-load pricing problem in the local telephone network and argue that the new electronic technology being brought on-line to provide local service reduces the cost of local usage and makes LMS inefficient.[7] It is now clear that the results of such benefit/cost tests are not robust and depend critically on the cost and demand elasticity parameters that are inputs into the benefit/cost models. Further, the welfare difference beween the two methods of pricing local service is not large.

The actual implementation of LMS, possibly because of the other changes that are sweeping the industry, has been slow and is now far behind the Bell operating companies' (BOCs') original implementation plans. Yet, local telephone companies continue to press regulators for LMS, albeit usually on an optional basis. Using demagoguery and a variety of noneconomic arguments, consumer groups largely oppose the introduction of LMS. More recently, they have seized on Park and Mitchell's results and argue that LMS should not be implemented because it is economically inefficient.[8]

Parallel with the debate over LMS, recent technological developments and institutional changes have brought about both the fact and the threat of local telephone competition. These have been detailed above. Simply put, technological change in switching and trunking has caused small-scale switching equipment, in the form of PBX-type switches, to move much closer to the end-user, reducing the minimum efficient scale of entry and allowing entrepreneurs to bite off portions of the traditional local network by concentrating traffic at such switches and connecting to the traditional network via high capacity trunks.[9] While some local telephone companies have resisted such local competition, it nevertheless has appeared and has been approved in most states, but with strings attached.

Competition and Flat-Rate Pricing

Where local telephone competition is present, any debate about the neoclassical benefits and costs of LMS is irrelevant. Beneficial or not on neoclassical grounds, strict, unsubsidized flat-rate pricing is unsustainable in a competitive environment. Ironically, in an unsubsidized environment, flat-rate pricing is sustainable only if subscriber usage can be measured so that a flat-rate tariff can be targeted and confined to a homogeneous group of users. In an unsubsidized, competitive local environment, beneficial or not in a neoclassical sense, one way or another LMS will become the dominant form of local telephone pricing. Incumbent local companies will have to adopt LMS or be largely driven from the local telephone market. Neoclassical benefit/cost tests, such as those done by Park and Mitchell, and Griffin and Mayor, are beside the point in today's competitive local telecommunications environment.[10]

The argument is very simple. Suppose that a neoclassical benefit/cost test such as the one performed by Park and Mitchell shows flat-rate pricing to be of net benefit. This means that, for the hypothetical *average* consumer, flat-rate pricing is of net benefit. But flat-rate pricing necessarily charges the same flat-rate price for all users, large or small. Two facts are well known: most subscribers have below-average usage, and under flat-rate pricing below-average users overpay and above-average users underpay, relative to the costs they cause. One of the fundamental principles of economics is that competition will be attracted by overpricing. Thus, under flat-rate pricing, if there are a sufficient number of below-average users who are overpaying, it will be possible for a competitor to capture these customers by offering them a measured service tariff. From the standpoint of the incumbent provider, losing subscribers who overpay relative to their costs will force the flat-rate price upward, thus exposing more of the local market to entry by measured service providers. The upshot will be that incumbents will be forced to adopt measured service themselves or be driven from the market.

Given that the scale of economical entry into the provision of local service is not likely to be large—the number of business lines connected to PBXs exceeded the number directly connected to local central offices in the early 1980s in the United States, and shared tenant services have begun to appear on a rather small scale in almost every state, even though local service is probably subsidized—it would appear that the flat-rate pricing of local service is not sustainable in an unsubsidized, competitive local environment.

It is tempting to suggest that the unsustainability of flat-rate pricing is not a long-run outcome, because other competitors could lure below-average users away from LMS by offering a flat-rate service that was not priced above the total cost of serving this subscriber. In order to win such a subscriber back from LMS with a compensatory flat-rate offering, such an offering would have to be targeted only to subscribers with usage appropriately below the average, and it is generally costly to identify and target such subscribers *ex ante*. If a second flat-rate offering is made that would lure below-average subscribers back, it would immediately be taken by *all* subscribers with usage above this level, thus rendering it noncompensatory to anyone who offered it.

Thus, with a dispersion of usage among subscribers, and with flat-rate pricing set on the basis of the average subscriber—as it must be if only one flat rate is charged—local competition will attract low-use subscribers with a compensatory LMS offering, even though LMS is not optimal in the neoclassical sense. In fact, if competition is allowed to proceed in this situation and the incumbent continually and correctly reprices his flat-rate service upward as competitors steal low-use subscribers, eventually flat-rate pricing will be forced out of existence.

The reason for this seemingly perverse result is simple. The LMS competitor can rely on self-selection by subscribers to win him subscribers that can be served on a compensatory basis. The flat-rate incumbent, or competitor, cannot. Only when he has a monopoly can a local service provider offer a sustainable flat-rate service.

Further Comments on Flat-Rate Pricing

There will be those who will use the above analysis to argue against local competition on the grounds that it may result in allocatively inefficient adoption of LMS. But the "nonoptimal" nature of the competitive outcome does not result from any failing of competition, but instead points up a failure of neoclassical welfare economics.

First of all, flat-rate pricing is nirvanic because there is no way it can be brought about effectively, in all dimensions of economic welfare, except by a perfect monopolist operating under perfect regulation. Such conditions are impossible under any known institutional environment. By allowing unregulated competition in local telephone markets, as we do in most of the other markets in the economy, as well as the LMS that will inevitably go with it, some theoretical and unachievable efficiency is traded for the discipline of better x-efficiency, better pricing efficiency, and better efficiency in the development and adoption of technological improvements.

Second, there are some deeper methodological issues here that should be mentioned. The fundamental problem with aggregate, neo-classical efficiency analysis is that it sums benefits and costs across disparate individuals and draws its conclusions based on a comparison of such aggregates. Yet, from the standpoint of individual decisionmakers, a comparison of individual benefits and costs may not produce the same result as the aggregate comparison does. Such individuals therefore are made worse off by a result that is of net benefit in the aggregate.

On the other hand, if we look at the world through the eyes of the individual, the standard of economic welfare is the voluntary exchange: any voluntary exchange represents an improvement in economic welfare because both parties are necessarily made better off or they would not voluntarily enter into the exchange.

In many cases there is no conflict between these two approaches to economic welfare. In a standard supply-and-demand framework, maximizing aggregate consumer and producer surpluses produces the same result as maximizing voluntary exchanges. In local telephone markets, it does not.

Thus, the fundamental flaw of the neoclassical aggregate approach is that it submerges the individual. As applied to benefit/cost analysis, it is collectivist. By aggregating, such benefit/cost tests assume at the outset that a monopoly exists to administer the flat-rate pricing. This is curious, given that the often stated goal of regulation is to produce a competitive outcome.

There are also some well-known equity problems with flat-rate pricing that are not present with the competitive, voluntary exchange view of economic welfare. Flat-rate pricing inevitably overprices some users and underprices others. Where all users are solidly on the network—meaning, the price elasticity of access to the network is effectively zero—this results in an inequitable redistribution of income from low to high users, and probably from low-income to high-income households. Where all users are not on the network, flat-rate pricing will keep some—probably low users—off the network because the costs of having a phone exceed the benefits. Thus, with any flat-rate tariff there are inequities, which are remedied by competition.

» *Further Thoughts and Problems*

The above remarks offer a way of looking at competition and the local exchange market, which is not to suggest that all of the answers to the problems facing the telecommunications industry are contained herein. What is suggested is that a realistic view of what is going on is needed before progress can be made. A solution cannot be devised unless the problem is understood. On this score, we have a long way to go.

Dealing with Ignorance

With all of the talk about deregulation of the telecommunications industry, it is ironic that the industry now chafes under the view of the U.S. District Court in the District of Columbia. As H.L. Mencken once said, "There is nothing so dangerous as active ignorance." The problem is that Judge Harold Greene's understanding of markets in general and the telecommunications market in particular is so fundamentally flawed that there is no conceivable way that the Modified Final Judgment (MFJ) will ever be modified in a sensible way as long as it is under his jurisdiction. It is useful to spell out these fundamental misunderstandings that now drive the court.[11]

1. The judge has never understood the role of the separations process in the events that have swept this industry in the past thirty years. Simply put, beginning with the Charleston plan in 1955 and culminating in the Ozark plan in 1970, the regulatory process allocated larger and larger portions of probably already inflated local revenue requirements to the interstate toll market. This process had two effects on competition, neither of which seem to be understood by the judge.

First, the overpricing of toll services attracted swarms of competition—from private microwave systems to the various specialized and other common carriers—to the toll market. It is debatable whether or not much of this competition would have appeared if the regulatory process had allowed toll to be priced at anywhere near incremental economic cost. As far as the interstate market is concerned, the real defendant in the various antitrust cases was not AT&T and the Bell System but the various separations plans.

Second, the underpricing of residence local service thwarted, and still thwarts, much local competition. In addition, with a large percentage of local costs "allocated" to the toll markets, and with the failure of the regulatory process to give local companies realistic depreciation rates, technological change and innovation in the provision of local service lagged. As argued above, the local network became "loop-heavy." Even though much local competition did appear in the form of private networks controlled by PBXs (which Judge Greene does not understand either [*U.S. v. Western Electric* 1987: 28]), much more local competition was thwarted, not because local service is a natural monopoly—as Judge Greene argues—but because much of the incumbents' local service was priced below both embedded and current cost.

Judge Greene's ignorance on this issue can be seen in the following: "[I]t is not at all clear that the subsidy . . . has ever existed. In its extended oversight of AT&T, and in investigations extending over many years, the Commission was never able to determine whether, in fact, local rates had been subsidized by long distance rates" (*U.S. v. Western Electric* 1983: 16). To which the following footnote was appended:

> To be sure, the Operating Companies have been receiving revenues from the Bell System under a division of revenue process, which allocated interstate toll revenues between the Operating Companies and Bell's Long Lines division. See 552 F. Supp. 196 n. 271. Whether this allocation was a "subsidy," that is, whether the Operating Companies received more than they earned or deserved, has never been proved. At the trial of this case, witnesses for the Bell System stated that there was such a subsidy; government witnesses contended that, in reality, the situation was the reverse: that local telephone revenues have been subsidizing AT&T's intercity rates. As the court has observed, "Since the trial was aborted by the settlement, no final decision was reached on this issue" 552 F. Supp. 169 n. 160 (*U.S. v. Western Electric* 1983: 16 n. 39).

To be sure, the DOJ bears some responsibility for the ignorance of Judge Greene on this point. To have sponsored a parade of witnesses who argued that local service was subsidizing toll was unconscionable.

2. Consider the following:

> Even if all state and local regulation prohibiting competitive entry into the local exchange market were to be repealed tomorrow, and any one were free, as a matter of law, to sell local telephone service, the exchange monopolies would still exist substantially in the same

form and to the same extent as they do now. The conditions that caused these monopolies to emerge in the first place—the need for connecting local customers of telephone service to the telephone company central office switches by means of wires strung or buried in millions of places throughout America's cities and rural areas, and the enormous capital resources required for this project—preclude any thought of a duplication of the local networks.

Only when a practical and economically-sound method is found for large-scale bypass or for connecting local consumers by a different method—as microwaves and satellites were ultimately found to be feasible for handling long distance traffic—can the Regional Companies' local monopoly be regarded as eroded (*U.S. v. Western Electric* 1987: 18 19).

This quote reveals four erroneous views of the local market:

1. The judge does not seem to understand the relationship between switching, trunking, and access lines in the provisioning of local service. Thus, he sees very little local competition because PBXs are apparently regarded as mere terminal equipment. As argued above, they are not terminal; such competition is an integral part of network competition.
2. He believes that local competition cannot exist unless the local network is duplicated. Ignoring for the moment the point that duplication may not be bad if it provides better price and cost discipline, the idea that parts of the network could be carved out by competitors using PBX-type switches is entirely missed.
3. The idea that potential competition could be a persuasive force in local competition, as it historically was, is not even considered, much less appreciated.
4. Finally, the judge shows no appreciation for the point that, aside from the legal prohibitions on local competition, the subsidization of residence service has thwarted much competition there.

Not only does Judge Greene miss most of the local competition that has occurred, but he presumes that his failure to find any is almost entirely because local service is a natural monopoly. This is simply not true.

3. *In simplest terms, competition exists when customers have sufficient actual or potential alternatives to provide discipline for sellers.*

The essence of competition lies in the actual or potential availability of alternatives that can be chosen by the consumer. The judge shows no appreciation of this. For him, on-the-spot alternatives in large numbers with small market share are necessary.

And even when such alternatives are available, the judge believes the consumer is always the captive of whatever choice is made! Consider the following: "Bypass is also unpalatable for many enterprises even if they could achieve it, because in bypassing the Regional Company bottleneck by connection of their equipment directly to an interexchange carrier, they would merely be exchanging one form of captivity for another" (*U.S. v. Western Electric* 1987: 30).

In Judge Greene's mind, with the customer always the captive of his choices, competition could never exist anytime, anywhere. If this is the view of competition that will be used to judge the course of the future of the industry, there is no hope for a deregulated and competitive telecommunications industry for the foreseeable future.[12]

The Huber and DOJ Reports

With all of the talk about the industry's problems, it is often forgotten that two very important features of the industry are still with us—separations and pervasive local regulation. And as previously suggested, these are responsible for most of the problems facing the industry today. What to do about them is not altogether clear, but at least these problems must be recognized before they can be addressed.

Thus, the tone of the Huber and DOJ reports in the recent triennial review of the MFJ line-of-business restrictions is encouraging. During the AT&T antitrust trial, the DOJ seemed to be doing its best to obscure the role of separations in the industry. While still not addressing the separations issue directly, the DOJ now at least seems to recognize that local regulation is a large stumbling block to effective local competition. And its proposal that the BOCs be kept out of the interLATA (local access and transport area) toll markets until local regulation was significantly eased certainly would have put useful political pressure exactly where it was needed. Further, the Huber Report shows an understanding of the real nature of the way in which the local network is evolving and the role that local competition can play in this evolution. The Huber Report articulated a badly needed long-run vision of the industry.

Implementing Reduced Regulation

The principle conclusion is that, in the long run, the pricing of local service can be largely left to the marketplace if local companies are not protected, handicapped, or subsidized.

In order to see how competition could be effective at the local level, consider a situation where local service was open to competition and stood on its own, and where two simple rules were in effect: (1) The local company was not allowed to discriminate among classes of users, meaning that the same price for access and usage would be charged for single and multiline business and residence and for PBX trunks. This presumes a measured local environment, probably with volume discounts, and does not preclude local price differentials based on, say, loop length or customer density. (2) The reselling of local service is allowed.

In such a situation, PBX-type competition would be poised to bite off portions of the network that were overpriced. In addition, PBX competition would ensure the competitive pricing of Centrex. Indeed, PBX competition already provides this discipline. With Centrex reselling allowed, this competition would in turn provide additional competitive discipline to single and multiline business and residence service. Where the reselling of Centrex is allowed, resellers have already successfully entered the single and multiline business market. If the incumbent company overpriced Centrex, it would lose this market to PBXs; if it underpriced Centrex, resellers would compete effectively against business and residence service. Competitive discipline would be virtually complete.

This set of conditions would allow virtually all telecommunications markets to be essentially deregulated. But given the differing interests of the various telecommunications carriers and the degree of coordination required from the various regulatory and legal authorities, the extent to which this has a chance of coming about is uncertain. If one of the participants has a veto over the above outcome, then such a fully deregulated scenario will be feasible only if all believe they are better off than under all other feasible outcomes.

There is a strong temptation by all concerned to use existing legal and regulatory processes to gain a competitive advantage. For example, the same political-regulatory arrangements that produced a regulatory mechanism that set up and enforced the popular toll-to-local subsidy is still there, and state regulators may want to keep local companies regulated and protected so that they can be used to continue the popular toll-to-local residence subsidy as much and as long as possible.

Another concern is that the perception of local companies may be that their self-interest requires staying under the protective wing of regulation as long as possible, at least until they are able to get back the depreciation that would be unfundable under competition. Since uneconomic depreciation schedules were mandated by regulators, this argument has some equity appeal. Further, local companies may want to get as much pricing flexibility and other freedom as soon as possible but still use the regulators to shield them from entry in both the traditional local service market and in the carrier access market.

Looking at only the local market, the possibility of deregulation probably depends on the relationship among: (a) the current cost of provisioning local residence service, (b) its embedded cost, and (c) the existing local residence prices. Most probably, at the present time embedded costs (revenue requirements) are above current costs, which are also above existing prices. As long as this situation exists, then deregulating local service in a competitive environment would result in an increase in local residence prices (to current marginal cost). Regulators would not find such a move attractive because it would be associated with higher local rates in the minds of their most powerful political constituency, the local residence subscriber. Even local companies might not be in favor of such an outcome; it might require them to write down some assets, unless this problem had been dealt with in the meantime.

If existing residence prices were at or above current costs, then regulators would be in favor of deregulation because it would not result in higher residence prices. If local companies had already recovered their unfunded depreciation, they might be willing to trade the resulting possibility of lower residence prices for pricing flexibility. If considerable unfunded depreciation remained, local companies would probably oppose such a move because it would again require them to write down some assets.

An environment most conducive to the implementation of local service deregulation would be one in which the local companies' capital recovery problem had been resolved and carrier access charges had been reduced to the vicinity of local transport costs. This is not likely to happen soon, and it is not clear what piecemeal efforts might be possible in the meantime. Any piecemeal movement—such as allowing the BOCs into the interLATA toll market, or the removal of the BOCs' one-plus intraLATA monopoly—runs the risk that those who benefit from such a move might try to stall the process at that point.

In sum, this analysis outlines a reasonable set of circumstances under which local competition can be substituted for local regulation. How we get from here to there is more problematic. One thing is certain, however: effectively dealing with the local companies' capital recovery problems and eliminating the inflated carrier access charges are the two prerequisites that would set the stage for the full deregulation of local service and would produce large gains in economic efficiency.

What to do in the meantime? While these preconditions for full local competition are not presently available, throwing the local market open to free and fair entry would be useful even if nothing else were done. At the very least, this would show very quickly where local prices are above cost and would force the regulatory process to deal with this distortion.

Many small local providers are concerned that a complete opening-up of the local market to competitive entry would result in so-called cream-skimming. But there is nothing wrong with cream-skimming. It is just a pejorative name for competition—competition brought about by the overpricing of some services. Cream-skimming gives the same corrective discipline as any competition and is therefore desirable.

It was emphasized above that it will take time for local competitive institutions to develop. Local competition has been suppressed for so long, by both the toll-to-local subsidy and the regulatory process, that local competitive institutions, such as Centrex resellers, have not had a chance to develop. A transition period along the lines of some of the "social contract" proposals might be useful to deal with this problem. Likewise, some modification of strict rate-of-return regulation would be useful.

» References

Alleman, James H. 1977. *The Pricing of Local Telephone Service.* Boulder, Colo.: U.S. Department of Commerce, Office of Telecommunications.
Baumol, William; John Panzar; and Robert Willig. 1982. *Contestable Markets and the Theory of Industrial Structure.* New York: Harcourt, Brace, Jovanovich.

Bonafield, Christine. 1986. "Local Measure[d] Service Planned in Widely Circulated Rand Study." *Communications Week* 104 (11 August): 8.

Bornholz, Robert, and David S. Evans. 1983. "The Early History of Competition in the Telephone Industry." In *Breaking Up Bell: Essays on Industrial Organization and Regulation,* edited by David S. Evans, pp. 7–40. New York: Elsevier Science Publishing Company.

Brock, Gerald W. 1981. *The Telecommunications Industry: The Dynamics of Market Structure.* Cambridge, Mass.: Harvard University Press.

Griffin, James M., and Thomas H. Mayor. 1987. "The Welfare Gain from Efficient Pricing of Local Telephone Service." *Journal of Law and Economics* 30, no. 2 (October): 465–488.

Haring, John. 1984. "Implications of Asymmetric Regulation for Competition Policy Analysis." Working Paper No. 14. Washington, D.C.: Federal Communications Commission, Office of Plans and Policy.

Huber, Peter W. 1987. *The Geodesic Network: 1987 Report on Competition in the Telephone Industry* (Huber Report). Washington, D.C.: U.S. Department of Justice, Antitrust Division.

Mathewson, G. Franklin, and G. David Quirin. 1972. "Metering Costs and Marginal Cost Pricing in Public Utilities." *Bell Journal of Economics and Management Science* 3, no. 1 (Spring): 335–339.

Mitchell, Bridger M. 1976. *Optimal Pricing of Local Telephone Service.* Publication R-1962-MF. Santa Monica, Calif.: RAND Corporation (November).

Mitchell, Bridger M. 1978. "Optimal Pricing of Local Telephone Service." *American Economic Review* 68, no. 4 (September): 517–537.

Park, Rolla Edward, and Bridger M. Mitchell. 1987. *Optimal Peak Load Pricing for Local Telephone Calls.* Publication R-3404-1-RC. Santa Monica, Calif.: RAND Corporation (March).

Schumpeter, Joseph A. 1950. *Capitalism, Socialism and Democracy.* 3d ed. New York: Harper and Brothers.

State Telephone Regulation Report. 1985. Vol. 3, no. 21. Alexandria, Va.: Telecom Publishing Group (7 November).

State Telephone Regulation Report. 1986a. Vol. 4, no. 16. Alexandria, Va.: Telecom Publishing Group (28 August).

State Telephone Regulation Report. 1986b. Vol. 4, no. 22. Alexandria, Va.: Telecom Publishing Group (17 November).

Telecommunications Reports. 1986. "RAND Study Says Local Measured Rates Cannot be Justified Over Flat Rates." Vol. 52, no. 33 (18 August).

U.S. v. Western Electric Company, et al. 1983. U.S. District Court for the District of Columbia, Civil Action No. 82-0192. *Opinion* (filed 20 April).

U.S. v. Western Electric Company, et al. 1987. U.S. District Court for the District of Columbia, Civil Action No. 82-0192. *Opinion* (filed 10 September).

Wenders, John T. 1987a. *The Economics of Telecommunications: Theory and Policy.* Cambridge, Mass.: Ballinger Publishing Company.
———. 1987b. "On Modifying the MFJ." *Telecommunications Policy* (September): 243–246.
———. 1988a. "On the Sustainability of Flat-Rate Telephone Pricing." Economic Discussion Paper No. 3 (September), University of Idaho, Department of Economics.
———. 1988b. "Local Telephone Competition." Economic Discussion Paper No. 7 (October), University of Idaho, Department of Economics.

» Notes

1. For further discussion of the role of competition in the telecommunications industry, see Wenders (1987a: chs. 2, 10, and 11).
2. While proper costs studies by individual firms may be useful from a firm's standpoint, the marginal costs that are relevant can probably only be uncovered by the competitive process. The market does much better cost studies than either accountants or economists.
3. For a more detailed discussion of these issues, see Wenders (1988b).
4. For useful surveys of the status of STS service, see *State Telephone Regulation Report* (1985: 1–6; 1986b: 1–8) and Huber (1987: ch. 1).
5. A more detailed version of Mitchell (1978) is Mitchell (1976).
6. See Wenders (1987a: chs. 2, 10, and 11) for a more extensive discussion of the benefits and costs of LMS.
7. See, for example, the reports in *Communications Week* (1986: 8), *State Telephone Regulation Report* (1986a: 8–9), and *Telecommunications Reports* (1986: 27).
8. See the report in *Communications Week* (1986: 8).
9. For useful surveys of the status of STS service, see *State Telephone Regulation Report* (1985: 1–6) and (1986b: 1–8).
10. For a more rigorous analysis of this logic, see Wenders (1988a).
11. For an elaboration of these points, see Wenders (1987b).
12. For further discussion, see Wenders (1987b).

15 » Local Exchange Pricing: Is There Any Hope?

William E. Taylor

» Introduction

The specter of competition in the local exchange business raises two important questions: (1) Can the local exchange remain a profitable venture if subsidy flows to the local loop disappear? and (2) Will traditional pricing structures remain viable against competitive offerings?

This paper analyzes pricing alternatives in the local exchange and concludes: (1) yes, and (2) no.[1]

Competition in the local exchange takes place in a setting constrained by regulatory concern and precedent. Average embedded loop cost (roughly $26 per month) may be too high for all customers to pay without threatening universal service. If the marginal cost (maybe $10 per month) were charged, efficient pricing would require that the remaining $16 be disproportionately borne by flat-rate subscriber access charges. This brings us back to universal service concerns. Thus, there appears to be a conflict between prices that support equity (universal service) and those that support efficiency.

What is required is a method to recover embedded loop costs efficiently within the local access market so that subsidies from relatively elastic services would be minimized. In addition, that method would have to be sustainable in a competitive environment so that current distinctions among services (such as local usage, toll usage, and interLATA [local access and transport area] carrier access) that are not based on costs cannot be maintained. A possible solution is local measured service (LMS)[2] in the form of a family of optional multipart tariffs designed to make all classes of customer, large and small, at least as well off as they were under the initial rates, while maximizing the profit of the firm.

This paper presents such a family of tariffs and shows in a simple example the mechanism by which they work. Problems of implementing such tariffs under competition and inflexible regulation are addressed, along with (1) pricing flexibility in social contracts, (2) consistency of local usage, toll usage, interLATA access and intraLATA access rates, and (3) pricing services in an intelligent ISDN (integrated services digital network) local exchange network.

Next, basic local exchange services, access, and usage are defined in economic rather than regulatory terms. Efficient pricing for local exchange access and usage is then examined, and it is shown how a set of optional multipart tariffs can be designed that (1) recovers the local exchange carrier's (LEC's) costs, (2) is viable under competitive conditions in the local exchange and intraLATA toll markets, and (3) makes all size classes of consumers at least as well off as they have been under traditional rate structures. Finally, implementation problems are pointed out: essentially, how far removed current rate structures and levels are from those contemplated in this paper, and how regulatory reforms such as price caps or social contract regulation may provide the required pricing flexibility.

» The Local Exchange Market

In recent years, market analysis of the potential profitability of the local exchange network has been schizophrenic. Some viewed divestiture and the access charge decision[3] as having created two separate and inherently unequal market opportunities:

1. A long-distance market that was profitable, technically innovative, responsive to market forces, and steadily growing

2. A local exchange network based on obsolete, unchanging technology and an imperfectly enforceable monopoly franchise, saddled by a public interest concern for universal service and financially sustainable only through subsidy flows from long distance

From this perspective, the only rational response on the part of LECs was to attempt to diversify out of the local exchange business and—among other enterprises—to wheedle their way into interexchange toll.

A simple examination of the structure of the local exchange market provides a sharp contrast to this view. Absent public interest and regulatory concerns, the profit potential from the imperfect monopoly that remains in some areas of the local exchange market appears to be enormous. In sectors of the market that are relatively secure monopolies (for example, residential access), demand is extremely inelastic at current prices, implying that profit could be increased through price increases. Moreover, the market is of significant size: local exchange is about the size of the toll market, as measured by revenue (see Figure 15–1), and while growing less rapidly than toll in recent years, local exchange is less subject to cyclical fluctuations (see Figure 15–2).

Subject to the whims and vagaries of regulation, rates—and rate structures—for local exchange services have evolved in very peculiar directions. Price levels have been kept artificially low by a number of standards. Current prices are probably a small fraction of the profit-maximizing prices for local services and are probably less than marginal cost for some services. Moreover, there is evidence that current prices are substantially lower than those that would be set by a profit-maximizing monopolist constrained by classic rate-of-return regulation. A very general result from microeconomic theory shows that a regulated firm that sets prices so as to maximize profit, subject to nearly any form of regulatory constraint, will operate only on the elastic portion of its demand curve: that is, where marginal revenue is positive and where the price elasticity of demand exceeds 1.0 in absolute value.[4] This is clearly not the case in most telecommunications markets.[5] The price level at which a service like residential access would become price-elastic is probably astronomical; indeed, it is difficult to think of a major telecommunications market in which demand appears to be elastic at current prices. One is left with the conclusion that prices of telecommuncations

FIGURE 15-1

» *The Telecommunications Marketplace (1985 sales: $145 billion).*

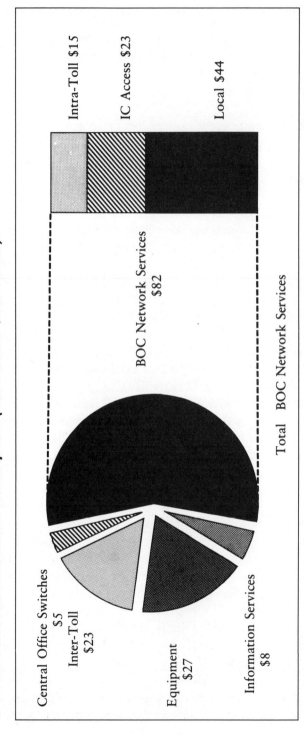

Source: Peter J. Huber, *The Geodesic Network: 1987 Report on Competition in the Telephone Industry* (Huber Report) (Washington, D.C.: Department of Justice, Antitrust Division, 1987), Tables G.7, G.10, L.2, and T.1.

FIGURE 15-2

» *Bell System Revenue, 1950–1981 ($ billions).*

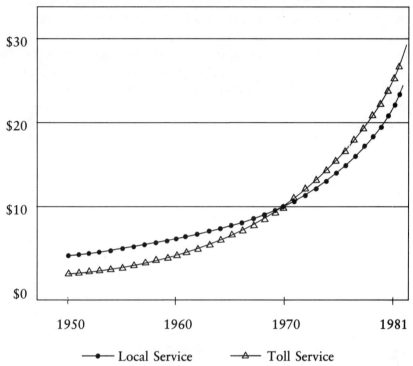

——•—— Local Service   ——△—— Toll Service

Source: FCC, *Statistics of Communications Common Carriers,* annual report, various issues.

services have been held below even the level that one would expect under classical rate-of-return regulation.

While telephone rates appear to be inexplicably low, the local exchange rate structure is equally inexplicable. The prevailing rate structure for residential users is currently a low flat-rate charge and generally no charge for usage. As fewer protected markets remain from which to subsidize residential access, the flat-rate charge will probably rise, possibly threatening universal service among low-income households. Although this rate structure is, and has been, ubiquitous, it is hardly the rate structure one would expect to see in a market in which there is a public interest concern for small users. Indeed, among the set of possible two-part tariffs,[6] it is the ideal tariff only for the customer class having the largest amount of usage.

Smaller users should prefer to pay a positive usage charge (per minute or per message) to reduce their flat-rate subscription fee.

The next section of this paper, examines rate structures that appear to make more sense in the local exchange. But first, since the intention is to price access and usage jointly in some clever way so that the local exchange can be a profitable business, we will be more explicit about what we mean by "access" and "usage."

In the abstract, we assume that the local exchange market provides two related services: customer access and customer usage. In the economics jargon, these services are complements (like shoes and shoelaces) because a decrease in the price of one will stimulate demand for both. "Access" is the service that has little value in itself but enables a customer to use the network—to place or receive calls—upon demand. The cost of access is the fixed cost (independent of usage) of installing and maintaining the customer's local loop.

"Local usage," on the other hand, is more complicated. Currently, local usage is defined as the service that provides a connection (switching and trunking) between customers in the local calling areas.[7] Think of dividing all use of the local network into intraexchange usage, interexchange local usage, intraLATA toll usage, and interLATA carrier access usage. Then regroup these calls, ignoring previous regulatory boundaries, into groups having similar cost characteristics—probably into groups of calls that differ by length of haul. Moreover, define the relevant mileage bands cumulatively so that a call in the third mileage band is also counted as a call in the first and second mileage bands. Draw the mileage bands sufficiently wide so that if we take "local usage" to mean all calls in the first mileage band, and "toll usage" to mean calls in the second and higher mileage bands, this does not distort the common definitions of "local" and "toll" too badly. In this nomenclature, an intraLATA toll call will consist of one local call (for the first distance band) plus additional toll segments, depending upon the distance of the call. An interLATA access call will consist of one local call (at a minimum) and additional toll segments, depending upon the location of the interLATA carrier's toll switch. If costs were independent of length of haul or number of times the call is switched, then all calls would be treated as local.

Note that it is the presence of competition in the local exchange that forces a rethinking of the standard definition of local usage. If there is no cost difference at the margin between a local call that connects two subscribers in the same exchange and an interLATA access call that connects a subscriber to an interexchange carrier (IC) point of presence (POP) in the same exchange, then competition will force the prices of those calls to be the same.[8] Thus, service distinctions based on jurisdictional and regulatory convenience and precedent that have no basis in costs cannot be maintained in the analysis.

» *Optimal Tariffs for Local Usage*

The problem is to price access to (and usage of) the local exchange network so that subsidies from other services are unnecessary and universal service is maintained. And if that were not difficult enough, the requirement is imposed that such tariffs be sustainable under competitive entry into the local exchange business.[9]

A simple solution to this generic problem was proposed many years ago: usage should be priced at marginal cost, and the firm's fixed costs should be recovered from subscribers in flat-rate access fees (Coase 1946: 169–89). Since usage price is equal to marginal cost, customers would face the correct incentives to determine their usage, and since the firm's fixed and variable costs are covered, the firm would earn a competitive rate of return. As long as the subscription fee induces no subscriber to discontinue service, economic welfare is the same as under pure marginal-cost pricing. Subscribers are assessed a lump-sum tax to cover the fixed costs of the firm, and provided consumer surplus from usage exceeds the tax, the firm's budget problem is solved through an infra-marginal transfer payment.

In the local exchange market, this solution is probably not feasible. Marginal cost of local usage is arguably close to zero,[10] and the fixed costs of the local loops are probably close to the $26 per month per subscriber. Hence, the Coase tariff would resemble current practice (in the absence of a subsidy from toll usage), in which the full cost of the local loop (plus the fixed portion of the cost of local switching and trunking) would be charged on a flat-rate basis to local subscribers and usage (local and toll use of the local loop) would be unmeasured and free. A local rate increase of this magnitude (probably more than 100 percent) would cause some marginal subscribers to disconnect and some households that would have joined the network to refrain from subscribing.[11] It would thus violate the infra-marginality assumption in the Coase result.

The Coase tariff, as it stands, is thus not feasible because it induces a small number of households to change their behavior regarding subscription to the network. However, if willingness to pay for telephone service differed across households and could be observed, a modified set of Coase tariffs could be constructed in which no customer was assessed a fixed charge in excess of his consumer surplus from usage. Marginal customers would be assessed relatively small fixed charges, and infra-marginal customers would be assessed relatively large fixed charges. These tariffs are similar in construction to Ramsey prices for multiproduct monopolies, in which the fixed costs of the firm are assessed relatively heavily to products that have relatively inelastic demand. In the current example, customers whose demand is relatively inelastic are bearing relatively more of the fixed costs of the firm. Welfare losses are minimized in both approaches, since welfare loss only occurs when an increase in price induces a decline in consumption.

The Theoretical Setting

Suppose initially that there are two consumer types, Big (B) and Little (L), with demand curves as shown in Figure 15–3; at any price, Big's demand $QB(P)$ exceeds Little's demand $QL(P)$. Initially, both customer types face a uniform price of P_0 per unit. The firm earns a normal profit:

$$[(P_0 - mc) * (QB_0 + QL_0)] - (\text{NTS cost}) = 0. \qquad (15.1)$$

Now give the two consumers a choice: let them continue to buy at the uniform price P_0, or buy under a two-part tariff (E_1, P_1) having a flat entry fee

$$E_1 = QB(P_0) * (P_0 - P_1) \qquad (15.2)$$

and a uniform usage charge

$$P_1, \text{ where } P_0 > P_1 > mc \qquad (15.3)$$

and E_1 is given by the sum of areas a and b in Figure 15–3.

Clearly, Little will stick with P_0; the increase in consumer surplus he would derive from the lower usage charge P_1 is given by the area a, which is more than offset by E_1, given in Figure 15–3 by the edged rectangle $(a + b)$. Big prefers the two-part tariff because E_1 takes only part of the increase in consumer surplus made

FIGURE 15–3

» *Welfare Gains from Two Optional Two-part Tariffs.*

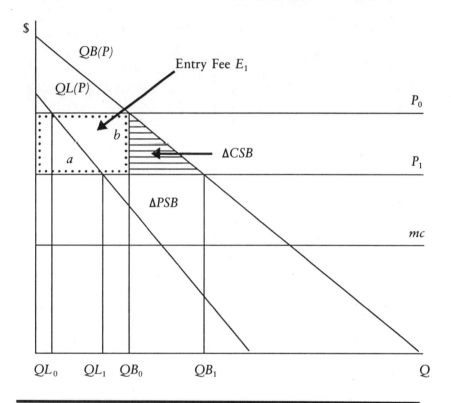

possible by the lower usage charge, leaving him better off by the lined triangle ΔCSB. The firm makes a non-negative profit from the two-part tariff because of demand stimulation induced by the lower marginal price: the increment $(QB_1 - QB_0)$ in Big's consumption takes place at price $P_1 > mc$. The rectangle ΔPSB represents increased producer surplus for the firm.[12] Thus, the firm as well as both customer types are better off facing the two-part tariff (E_1, P_1) than facing the uniform price P_0. Such a tariff is said to "Pareto-dominate" the flat-rate tariff P_0.

As we see in Figure 15–4, what we have done is equivalent to constructing a declining block tariff with usage charges P_0 and P_1 and a break point at QB_0. The result is true in general: a declining

FIGURE 15–4

» ***The Equivalent Declining-block Tariff
   for Two Customer Types.***

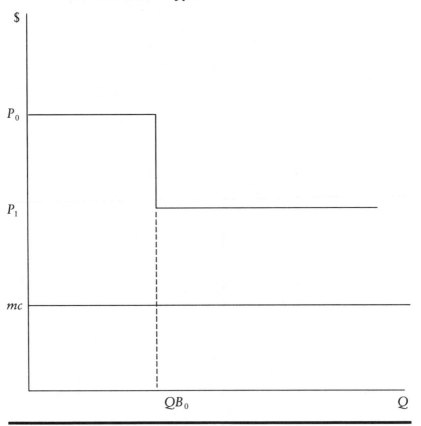

block tariff can always be viewed as the lower envelope of a set of
two-part tariffs from which consumers select their optimal consumption points (Faulhaber and Panzar 1977). On that declining block
tariff, the two consumer types select consumption levels QL_0 and
QB_1.

Now suppose that there are three consumer types: Big, Medium
(M), and Little. Their demand curves are consistent with the non-
crossing assumption and are shown in Figure 15–5. We can construct a set of optional two-part tariffs—(E_1,P_1), (E_2,P_2)—in the
same way we did above:

$$mc < P_2 < P_1 < P_0$$

$$E_1 = QM_0 * (P_0 - P_1)$$

$$E_2 = QB_0 * (P_0 - P_2) \qquad (15.4)$$

Using the same arguments as before, it is true that if Medium took (E_1,P_1) and Big took (E_2,P_2), then each would be better off than under P_0, and the firm would make higher profits. However, there is a potential complication: Big might prefer (E_1,P_1) to (E_2,P_2). If he did, then the firm might make less profit from him under the optional tariff than under the flat rate E_0, so that the firm might see higher profits under the flat-rate tariff.[13] To ensure that the set of optional two-part tariffs Pareto-dominates the flat-rate

FIGURE 15–5

» *Welfare Gains from Three Optional Two-part Tariffs.*

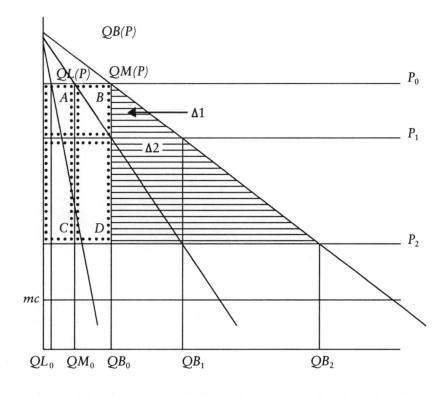

tariff, we have to further constrain P_1 so that the lower entry fee that Big would pay under (E_1, P_1) is offset—for his demanded quantity—by the higher usage charge. The decrease in consumer surplus that Big would undergo under (E_1, P_1), owing to the fact that $P_1 > P_2$, is given by the area of the trapezoid $(C + D + \Delta 2)$ in Figure 15–5. The reduction in the entry fee is given by the figure $(B+C+D)$. Thus, if $B < \Delta 2$, Big will prefer (E_2, P_2) to (E_1, P_1). We refer to this added complication as the "incentive compatibility constraint."

Note that by making P_1 suitably close to P_0, we can always ensure that this constraint will be met. As P_1 approaches P_0, $\Delta 2$ rises, while B declines. Thus, at some level of P_1 the increased usage charge in going from P_2 to P_1 more than balances the reduction in the entry fee, inducing Big to select the tariff (E_2, P_2) that assures the firm of higher profits from the optional two-part tariffs.

Summarizing the three consumer–type argument, if

$$mc < P_2 < P_1 < P_0$$

$$E_1 = QM_0 * (P_0 - P_1)$$

$$E_2 = QB_0 * (P_0 - P_2) \tag{15.5}$$

and the incentive compatibility constraint

$$B = (QB_0 - QM_0) * (P_0 - P_1) \leq \Delta 2$$

$$= (P_1 - P_2) * [QB_2 + QB_1 - 2QB_0]/2 \tag{15.6}$$

is met, then the following happens:

- Big chooses (E_2, P_2);
- Medium chooses either (E_1, P_1) or (E_2, P_2);[14]
- Little stays at P_0; and
- the firm makes higher profits than at P_0

Relative to the flat-rate tariff P_0, we refer to the set of optional two-part tariffs $\{E_i, P_i\}$ $(i = 1, 2)$ as Pareto-dominating and incentive-compatible (PDIC).

As shown in Figure 15–6, allowing consumers to choose among the set of three optional two-part tariffs—$(0, P_0)$, (E_1, P_1), and (E_2, P_2)—is equivalent to presenting them with a declining block tariff with usage charges P_0, P_1, and P_2 and break points QM_0 and QB_0. Proceeding in much the same fashion, with N consumer types one could construct $N-1$ optional two-part tariffs, with the result that no economic agent would be worse off than under the flat rate P_0, and some (including the firm) would be better off. This set of $N-1$ two-part tariffs would then be equivalent to a particular N-part declining block tariff, which would Pareto-dominate the flat-rate tariff P_0.

Competitive Optional Tapered Tariffs

The optional two-part tariffs can Pareto-dominate a uniform tariff, and by extension, a set of N optional two-part tariffs can

FIGURE 15–6

» *The Equivalent Declining-block Tariff for Three Customer Types.*

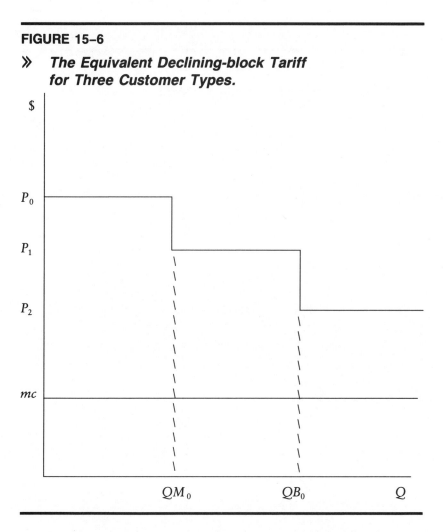

Pareto-dominate a set of $N-1$ optional two-part tariffs when there are at least N different customer types. This raises the question of how the optimal number of tariff steps is determined. In particular, if the market is competitive, how complex will the equilibrium price structure be?

An alternative representation of these optional two-part tariffs is given in Figure 15–7, where expenditure is shown on the vertical axis and usage on the horizontal. The slope of the expenditure lines is given by the prices $(P_0,P_1,P_2,0)$ and the entry fees are $(0,E_1,E_2,E_3)$. The hatched lower envelope of those expenditure lines is the set of undominated tariffs: that is, those tariffs for which there exists a customer type (a range of usage) for whom the given tariff is optimal.

FIGURE 15–7

»   *A Set of Optional Two-part Tariffs.*

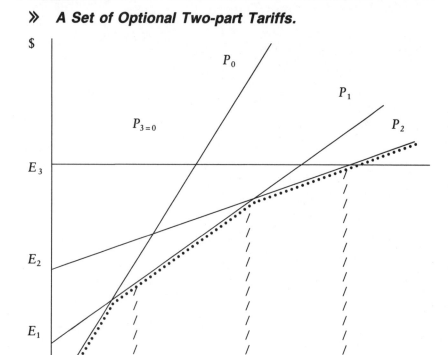

Note first that only undominated tariffs will be chosen by customers. If tariff choice is optional, only undominated tariffs will ultimately be offered; therefore, only undominated tariffs need be considered in this analysis. Second, it is interesting to see where the current rate structure for local usage fits into Figure 15–7:

- The rate structure for local usage (as defined in current tariffs) is given by (E_3,P_3) and is clearly optimal for only the very largest customer type.[15]
- Toll and interLATA IC access services are priced according to $(0,P_0)$, which—oddly—is desired by only the smallest customer type.

Given the regulatory history of the local exchange, one would expect that its services would be priced in a way that favored small local users at the expense of toll and IC access users. Paradoxically, existing price structures for local usage favor large customers, and for toll and IC access usage, favor small customers. It seems to be backwards.

Recall that local usage (as defined in this paper) is the local component of all types of usage, including local, intraLATA toll, and interLATA IC access. In a competitive local exchange, the current plan for pricing local exchange services thus:

- charges different rate levels for calls that have similar cost characteristics; and
- imposes different rate structures for otherwise identical calls—indeed, rate structures that favor only the very smallest or the very largest customers.

It should be clear that such tariffs will not be sustainable under competitive conditions.

Examination of Figure 15–7 also helps to illuminate the type of tariff structure that would emerge in a competitive local exchange market. If an LEC persists in offering only the tariff (E_3,P_3), a competitor could enter the industry with tariffs (E_i,P_i), $i=0,1,2$ and attract all customers but the very largest. Even if the competitor has higher fixed and marginal costs than the incumbent, entry is possible provided there are enough small, medium, and large customers that the entrant's markup over variable cost exceeds his fixed cost. In general, entry will be successful if there are enough customers who can be attracted away from the set of tariffs offered by the incumbent to cover the fixed costs of the entrant.

This analysis suggests an equilibrium set of optional two-part tariffs, customized to every large niche of customer usage types. Only transaction costs and the homogeneity of the customer population would limit the number of such tariffs that would emerge.

Consumer Welfare Under Tapered Tariffs: An Example

Assume a simple demand model that relates local usage demand for a customer of type i to the full price of the call to the user. Thus $Q_i = Q_i(r+P)$, where r is the incremental price charged for IC interLATA service or LEC intraLATA service, and P is the local usage charge.[16] If the call is local (by current definition), $r = 0$. If the call is toll, then r is the IC or LEC toll rate, net of access charges. If one takes the view that access is an input in the production of long-distance service, and if one also assumes that price is equal to marginal cost in the IC market, then changes in consumer surplus for customer i owing to changes in P can be calculated by writing $r = r(P)$ in equilibrium and integrating under the equilibrium input demand curve. Thus, if P is reduced from P' to P'', the effect on consumer i is given by[17]

$$\Delta CS_i = \int_{P'}^{P''} Q_i[r(P) + P]dP \qquad (15.7)$$

Thus, to compare the change in consumer surplus in moving from a uniform to a tapered tariff, all we have to do is calculate the area under each consumer type's demand curve above the marginal price he faces, subtract his entry fee, and subtract his consumer surplus under the flat-rate tariff.

For convenience sake, assume that the equilibrium input demand function for consumer type i can be approximated by a simple isoelastic form

$$\log(Q_i) = \log T_i - e_i \log(P_i) \qquad (15.8)$$

where T_i is a taste parameter and e_i is the price elasticity of demand for a customer of type i. We assume there are six different customer types having a hypothetical local usage distribution, along with average monthly usage for customers in each band, as shown in Table 15–1. This usage was assumed to have been generated by a uniform price of approximately \$.038 per minute.

» **Table 15–1.** **Distribution of Local Usage and Elasticities.**

Usage Band	Percentage of Accounts	Monthly Minutes of Use	Price Elasticity
0–120	74.03	29.10	−0.16
121–2,000	25.47	320.42	−0.16
2,001–4,000	0.26	2,728.92	−0.50
4,001–14,000	0.17	7,095.54	−0.50
14,001–40,000	0.05	22,052.14	−0.70
40,000+	0.02	134,851.20	−0.98

From the six usage bands, we construct six user types, with taste parameters T_i given by

$$T_i = [\text{avg mou}]_i\,[.0378]^{e_i} \qquad (15.9)$$

For the present purpose, judgmental estimates of price elasticities of demand are used for two reasons:

- Local usage includes the local component of toll calls as well as completed local calls. Without further information about the mix, it is difficult to estimate an appropriately weighted average price elasticity of demand.
- Price elasticities of demand by customer size class are difficult to find.

Assume a marginal cost of access for each consumer type of about half a cent per minute, which represents an average of peak and off-peak marginal costs. Note that this may overstate the true marginal cost for toll access usage for the largest customer segment, since switched access may not be the most efficient form of access for some of these very large customers.

We now compute sets of optimal two-part tariffs that maximize firm profits, subject to the incentive compatibility constraints described above. First, each two-part tariff (E_i,P_i) must Pareto-dominate the flat-rate tariff of $0.0378 per minute. This is guaranteed by choosing

$$E_i = Q_i(\$0.0378) * [.0378 - P_i] \qquad (15.10)$$

Second, given that (E_i,P_i) is designed to Pareto-dominate the flat rate in this way, each customer type i must find it optimal to actually select this tariff and not some other tariff (E_j,P_j), where $i \neq j$. In the isoelastic case, these two constraints are written:

$$C_1: \; E_i = T_i * (.0378)^{-e_i} * (.0378 - P_i) \qquad (15.11)$$

$$C_2: \; T_i/(1 - e_i) * [(P_{i-1})^{1-e_i} - (P_i)^{1-e_i}] > E_i - E_{i-1}$$

Hence, the optimal set of optional two-part tariffs—the optimal taper—is generated by solving the following nonlinear programming problem: Maximize with respect to $\{E_i, P_i\}$, the sum from $i = 1, \ldots, 6$ of

$$[E_i + (P_i - .005) \; T_i \; P_i^{-e_i}] \; d_i \qquad (15.12)$$

(where d_i is the demand share of the i'th type), subject to C_1 and C_2.

For the assumed elasticities, the results of the optimization are given in Table 15–2. The smallest two consumer types purchase from two-part tariffs that are indistinguishable from the uniform charge of $0.0378. Consumer types 5 and 6 receive deep discounts on the usage charge, in exchange for substantial entry fees. Recall that a characteristic of this tariff is that profit earned from each consumer type is no lower than it is under the flat-rate tariff. Overall profit under this tariff is 9.5 percent higher than under the uniform rate of $0.0378 per minute. Consumer surplus gains range from substantial for large users to zero for customer types 1 and 2. Consumer surplus per capita rises by $0.50 per month.

Thus the profit and welfare gains to the LEC and to consumers can be substantial relative to those available under a single uniform tariff. In the context of local exchange pricing, this example suggests that there are welfare gains to be had in moving from a single price for local usage (which includes the local component of toll usage) to a set of two-part tariffs in which large users pay higher flat monthly charges in return for lower marginal usage prices.

This example has some obvious limitations:

1. The usage and marginal cost data are essentially hypothetical and not very realistic.
2. The welfare and profit gains ignore effects of competition.
3. The welfare gain, as calculated, ignores the fact that large customers are predominantly businesses that respond to a reduction in toll rates by increasing usage, increasing production, and lowering prices of their products and services. When we calculate welfare gains of small customers above, this important effect is neglected.

» **Table 15-2. Optimization Results
(changes relative to uniform $0.0378).**

Customer Type	Entry Fee ($/month)	Usage Charge ($/min)	Change in Consumer Surplus	Change in Profits
1	~ $0	$0.0378	$0	$0
2	~ 0	0.0378	0	0
3	0.52	0.0376	~ 0	0.23
4	29.18	0.0337	0.82	12.06
5	342.17	0.0223	66.0	170.61
6	3,495.58	0.0119	2,350.0	1,956.60

Change in aggregate profit = $0.50 per customer per month
Change in consumer surplus = $0.50 per customer per month
Change in total welfare = $1.00 per customer per month

4. Finally, these calculations depend upon estimates of demand equations for customers of different sizes, and these estimates are surely imprecise. However, varying the elasticities significantly has no effect on the basic conclusions:

• that lowering usage rates for large users relative to small users can produce significant welfare gains; and
• optional two-part tariffs can be designed to benefit some size classes and harm none.

» *Implementing More Efficient Local Exchange Tariffs*

Pessimistic analysts of the local exchange business observe tariffs such as $(E_3, 0)$ in Figure 15–7, note that E_3 must rise as subsidies from other markets phase down, and conclude that either:

1. universal service is incompatible with competition and efficient pricing, so that society must forgo one or the other, or
2. other markets must be found in which the LEC has sufficient market power to raise contributions to cover the fixed costs of local exchange plant.

Neither of these alternatives is particularly appealing. Fortunately, measured service in the local exchange makes it possible to distinguish customers with different degrees of willingness to pay for telephone usage, and efficient optional two-part tariffs can be designed that make no users worse off, make some users better off, and make more profit for the firm.[18]

Three issues are treated below that arise in the implementation of this type of solution to the local exchange pricing problem: (1) the economic efficiency of local measured service relative to flat-rate service, (2) compatibility of optional two-part tariffs with current prices and conditions, and (3) possible regulatory settings in which rate structures such as those described above can occur.

LMS and Economic Efficiency

The classic test of the relative efficiency of LMS compares the savings in capacity costs under LMS with the added cost of measurement and nets out welfare gains and losses from additional calling and from calls forgone because of higher peak-period prices and because of changes in the value of calls that are blocked at the peak. Attempts to measure the net benefit from shifting an exchange to LMS have obtained different results, largely stemming from different assumptions (or conclusions) about the values of key parameters.[19] One might conclude from the debate that the welfare gain, if positive, is not compellingly large.

A problem with these studies is that they all assume that a uniform price will be charged all customers at any time of day. However, the introduction of LMS makes possible more than just efficiency gains from load-leveling. As outlined above, LMS makes it possible to distinguish customers with high and low willingness to pay for service, by means of customer self-selection of optional tariffs. Welfare gains from the ability to distinguish demand differences across customers must be added to the benefits accruing from implementation of LMS.

Absent measurement of usage, it would be difficult to observe a dimension along which customer willingness to pay could be distinguished. A possibility might be the number of access lines demanded per customer, and a tapered rate schedule could be developed along the lines outlined above with respect to the flat-rate charge per line. However, since the demand for lines is thought to be price-inelastic relative to the demand for usage, the welfare gains from such tariffs are likely to be significantly smaller than in the usage case. The primary determinant of the welfare gain from optional two-part tariffs is the price elasticity of demand of large users, since it is the stimulation from reducing their marginal prices that creates both welfare and profit gains.

» *Compatibility with Current Rates and Procedures*

As competition enters the local exchange market, problems with a number of rates, rate structures, and ways of doing business will become evident. Three such issues that impact the possible implementation of optional two-part tariffs are;

1. Rate disparities among types of local usage
2. Principles of equal access and open network architecture (ONA) applied to local service
3. The LEC's present and future ability to tariff services rather than use of the network

First, recall that local usage is defined in this paper as the local component of calls that are currently defined as local, intraLATA toll, or interLATA IC access. Today, local calls are generally priced somewhat below marginal cost (at zero), the local component of LEC toll calls is bundled in the intraLATA toll rate, and the local component of IC access is priced far in excess of marginal cost (in both the federal and state jurisdictions). Since each of these calls makes identical use of the local exchange network (that is, imposes identical incremental costs on the exchange), it will not be possible to price them substantially differently in a competitive market. The consequences of this observation are disconcerting. Current contribution flows within the local exchange from toll and IC access to "pure" local usage and residential access will disappear along with rate disparities between access and toll.

When competition comes to the local exchange, it will presumably come first to the toll market.[20] A consequence of competition will be the requirement to tariff access to the intraLATA toll switch so that current ICs can compete on an equal footing with the LEC.[21] This will be difficult to achieve without significant shifts in rates: current intrastate IC access charges (as well as interstate) are generally very high, relative to the rate for short-haul intraLATA toll. Hence, if an LEC were to impute an intraLATA access charge to itself that was commensurate with its current interLATA access charges, it would have to increase significantly the intraLATA toll rates. More likely, of course, would be that access charges and short-haul toll rates would have to be reduced, to the extent that such calls have similar costs to ordinary local traffic.

Second, competition in the local exchange market will undoubtedly require that some provisions be made to equalize the terms under which a customer can access all potential carriers. And it will most likely be the case that all of the intraLATA usage market will be subject to competition. There is only one distinction possible on the basis of cost: presumably calls within an exchange would have to be served by the access line provider, whereas all calls between exchanges could be competitively provided.[22] Thus, tariffed access rates would have to be established for all interexchange calls, not just for traffic that is currently classified as intraLATA toll.

From this perspective, the interLATA equal access arrangements may not be appropriate for the intraLATA toll market. Customers may know to dial "1" for long distance (presumably a different area code), but few customers understand LATA boundaries, and even fewer can distinguish a distant intraexchange call from a toll call. If all carriers (LECs and competitors) were allowed to serve all markets, equal access would be simple—at least in principle. Dial tone would be obtained from the provider of the local loop, but all usage would be defaulted to a presubscribed carrier. The customer would not have to know his call's jurisdiction in order to dial—unless he wished to presubscribe to different carriers for different jurisdictions. However, if the LEC is forbidden to handle interLATA traffic, the customer must presubscribe to two (possibly identical) carriers (one for interLATA and one for intraLATA traffic) and must be able to distinguish interLATA from intraLATA calls. Equal access in the local exchange is thought by some regulators to be a necessary condition for allowing former Bell System LECs into the interLATA market, but implementation considerations suggest that provision of equal intraLATA access and removal of interLATA prohibitions must be done concurrently.

Third, most current tariffs are service-oriented rather than cost-oriented. Thus, local usage for IC access is tariffed very differently from intraexchange usage, and special access facilities provided for data are tariffed differently from those provided for video. As pointed out above, competition and the ensuing arbitrage will make such distinctions difficult; indeed, they can be maintained only in markets in which the LEC retains some market power. However, new directions in network evolution suggest that even in those markets the LEC will be unable to tariff different services differently.

The intelligent ISDN network of the future will provide functionality rather than services to customers. Functions such as switching and transmission of a given bandwidth at a particular rate will be provided by the LEC, and the customer will use network peripherals or customer-premise equipment (CPE) to construct whatever final service he requires. Since services are under the control of the end-user, and since the LEC cannot tell what service is actually being provided, prices will have to be based on the underlying cost characteristics of the network.

Pricing Flexibility and Ceiling Price Regulation

It is apparent that competition will emerge in the intraLATA market while the incumbent LECs are still subject to pervasive regulation. There is thus a further requirement for the proposed optional two-part tariffs. They must not only be PDIC and sustainable against competition, but they must also be appropriate in a regulated context.

At first glance, tapered usage charges have not been an unmitigated success in regulatory hearings. This paper has made no mention of a cost justification for tapered usage charges,[23] and ignoring costs seems to contradict recent regulatory and legislative actions that discourage declining block tariffs (primarily in electricity) without an argument that costs decline with the amount of usage.[24] However, these actions have generally been based on the belief that charging a large customer less than a small customer will induce the large customer to inefficiently overconsume relative to the small customer. Now the marginal cost for local usage appears to be much lower than any rate that anyone is proposing to charge for any level of usage. Thus, proposed tapers do not induce large users to consume an inefficiently large amount of local usage, since they face a price that is still in excess of marginal cost. The perspective of this paper is that some contribution over marginal cost must be raised from usage charges. The task is to find the most efficient tariff that will accomplish this goal.

A second issue is that considerable pricing flexibility would be required to implement successfully a set of optional two-part tariffs. Little enough is known about demand characteristics of customers under different rates and rate structures; accurately forecasting "take-rates" of particular optional tariffs would be difficult. Current tariff filing requirements (cost support, notice, hearings, and so on) would make optional tariff proposals very risky. Also, rapid action would be necessary to respond to competitive tariffs designed to attract customers in particular size niches.

Fortunately, looming on the horizon is a regulatory reform that provides precisely the degree of flexibility that would be required. Price cap regulation in the federal jurisdiction and "social contract" regulation in the state jurisdiction appear to be natural settings for implementing optional two-part tariffs for use of the local exchange. The focus of both of these regulatory reforms is an index of service prices; provided a weighted average of LEC prices remains below the established ceiling, the LEC is free to set individual service prices moderately freely. The Further Notice of Proposed Rulemaking (1988) implementing price caps for AT&T proposes that streamlined regulation be applied to price changes consistent with the cap. If this standard were applied in both federal and state jurisdictions to the LECs, optional two-part tariffs could be implemented immediately.

Recall that the construction of the optimal tariffs designed above effectively reduces the marginal usage charge to all customers[25] in exchange for fixed entry fees. Since the original uniform tariff is always available, no customer is made worse off by the existence of additional optional tariffs, and streamlined regulatory oversight would be adequate protection from market power in the local exchange. In addition, offering optional, two-part, Pareto-dominating tariffs will have the desirable effect of lowering the average price of the price-capped firm.

» Summary

Local exchange services—broadly viewed to include access and all usage of the local network—can be a profitable business in the absence of subsidies from other services. Key to this conclusion is the realization that traditional service definitions, rate levels, and rate structures will not be viable in a competitive environment. There exists a class of rate structures (Pareto-dominating, incentive-compatible, optional, two-part tariffs) that maximizes firm profit while ensuring that no customer is worse off than under a uniform tariff. If the local exchange market is viable,[26] tariffs from this class will be profitable for the firm.

» References

Anderson, James E. 1976. "The Social Cost of Input Distortions: A Comment and a Generalization." *American Economic Review* 66, no. 1 (March): 235–238.

Bailey, Elizabeth E. 1973. *Economic Theory of Regulatory Constraint.* Lexington, Mass.: D.C. Heath and Company.

Brown, Stephen J., and David S. Sibley. 1986. *The Theory of Public Utility Pricing.* New York: Cambridge University Press.

Coase, Ronald H. 1946. "The Marginal Cost Controversy." *Economica* 13: 169–189.

Faulhaber, Gerald R. and John C. Panzar. 1977. "Optimal Self-Selecting Two Part Tariffs." Bell Laboratories Economics Discussion Paper No. 77 (January).

Federal Communications Commission. 1986. "Memorandum Opinion and Order in the Matter of the Annual 1987 Access Tariff Filings" (released 24 December).

———. 1988. "Further Notice of Proposed Rulemaking in the Matter of Policy and Rates Concerning Rates for Dominant Carriers." CC Docket No. 87–313, FCC 88–172 (adopted 23 May 1988).

Griffin, James M., and Thomas H. Mayor. 1987. "The Welfare Gain from Efficient Pricing of Local Telephone Service." *Journal of Law and Economics* 30, no. 2 (October): 465–487.

Park, Rolla Edward, and Bridger M. Mitchell. 1987. *Optimal Peak-Load Pricing for Local Telephone Calls.* Publication R–3404–1–RC. Santa Monica, Calif.: RAND Corporation (March).

Perl, Lewis J. 1983. *The Residential Demand for Telephone Service.* White Plains, N.Y.: National Economic Research Associates (December).

Schmalensee, Richard. 1976. "Another Look at the Social Valuation of Input Price Changes." *American Economic Review* 66, no. 1 (March): 239–243.

State Telephone Regulation Report. 1987. Vol. 5, no. 18. Alexandria, Va.: Telecom Publishing Group (24 September).

Willig, Robert D. 1978. "Pareto-Superior Nonlinear Outlay Schedules." *Bell Journal of Economics* 9, no. 1 (Spring): 56–69.

» Notes

1. Opinions in this paper are those of its author and not necessarily those of his employers. Some of these results derive from previous work on access charges performed jointly with David S. Sibley, who bears no responsibility for this manifestation of those ideas.

2. Where measured usage is more broadly defined so as to include all usage having similar cost characteristics.

3. CC Docket 78–72, in which the connection between interexchange carriers and the local exchange was changed from a system of transfer prices internal to the Bell System to a set of tariffed rates between independent LECs and interexchange carriers.

4. These results do not depend on the exact form of the regulatory constraint and include classical rate-of-return regulation; see, for example, Bailey (1973: 31, Proposition 3.5).

5. Elasticities in telecommunications range from roughly -0.03 for residential access (see Perl 1983) to -0.68 for interstate switched services (see FCC 1986: 1, Chart A).

6. Tariffs having a fixed and variable component.

7. Obviously, the distinction between local usage and short-haul toll usage is arbitrary. Currently, it is determined by regulatory boundaries that differ substantially across jurisdictions.

8. Of course, there may be cost differences between calls to a customer and calls to an IC POP, for example, because of possible economies of scale in point-to-point transport. For simplicity, this possibility is ignored without passing on its merits.

9. Recall that when discussing pricing of local usage, the use that interLATA and intraLATA toll make of local exchange facilities is included. Problems of undoing the current distinction between toll and local use of the local network will be discussed in the section entitled, "Implementing More Efficient Local Exchange Tariffs."

10. Averaged over usage at peak and nonpeak periods.

11. For both reasons, the network would be smaller by roughly 3 percent of the subscriber base, extrapolating from current econometric results, for example, Perl (1983).

12. This is a simplified version of a more general result that appears in Willig (1978).

13. For an example of this, see Brown and Sibley (1986: 87).

14. If Medium takes (E_2,P_2) instead of (E_1,P_1), he is—by definition—better off and the firm must necessarily make more money than if he remained on (E_1,P_1).

15. More accurately, $(E_{3,0})$—with E_3 substantially greater than marginal cost—is the local rate structure towards which the industry is evolving as subsidies from toll are phased down. Historically, the local rate structure has been a line parallel to $(E_{3,0})$ but much lower, probably in the neighborhood of marginal cost.

16. This setup is easiest to understand if IC access charges are billed directly to the end-user, although billing plays no role in the analysis. Note also that intraLATA toll calls are assumed to require access charges to cover use of the local exchange network.

17. See Anderson (1976: 235–38), or Schmallensee (1976: 239–43).

18. In a rate-of-return regulated setting, this additional profit would be returned to customers by adjusting the entry fees.

19. See, for example, Park and Mitchell (1987), which finds LMS to be inefficient, or see Griffin and Mayor (1987), which reaches the opposite conclusion.

20. Indeed, some thirty-nine states have already permitted some form of intraLATA toll competition; see *State Telephone Regulation Report* (1987: 3).

21. Note that an IC's intraLATA toll switch is presumably its POP, and the LEC's toll switch is presumably co-located with its end-office switch or tandem.

22. Competitors in the intraexchange market would have to overcome the handicap of requiring additional switching and transport beyond that required by the LEC. Nonetheless, depending upon how access is provided for intraLATA calls, competitors may find it profitable to provide intraexchange service on a competitive footing as part of a total telephone usage package.

23. There is probably general agreement that the marginal cost of switched usage is roughly invariant to the amount of usage purchased.

24. Recall that the Public Utility Regulatory Policies Act of 1978 (PURPA) recommended the elimination of tapered tariffs unless they reflect the cost of service.

25. Or at least does not increase it to any customer.

26. That is, if the fixed costs of the local network can be recovered from that network without unacceptable harm to universal service.

16 » *Pricing Local Exchange Services: A Futuristic View*

Bridger M. Mitchell

There is a fourth dimension to telephone pricing—*time*. To explore this dimension from an economic perspective, this paper will first review the sources of costs in a telephone network and how costs relate to pricing. Networks are exceedingly capital-intensive; costs are primarily due to investment in network capacity, and network utilization varies greatly. Time-of-day rates are one possible method of managing loads and improving capacity utilization, but their effectiveness is necessarily limited.

A promising method of making more efficient use of the network is a new type of rate structure—dynamic pricing—in which the actual prices of network services would vary more or less continuously as the network load changes. This paper will examine the possibility of real-time pricing of the local network.

» *Telephone Network Pricing and Costs*

Prices

Prices have four key roles in the telephone industry, as they do in any sector of the economy.

First, prices raise revenue for suppliers. In general, revenue from prices must be sufficient, in sum, to cover the costs of supply, something that is taken for granted in a market-oriented economy. However, in countries where the telecommunications enterprise may be in the public sector, prices in aggregate may have to cover either less, or more, than the total cost of supply.

Second, prices reduce consumers' demand for service and act to "allocate" the available services among potential customers.

Third, prices are a yardstick of value. They measure the minimum benefit consumers derive from telephone service. Subscribers, voting with their dollars, are saying that they willingly give up spending income on other goods in order to make telephone calls at current prices. And on the supply side of the market, telephone companies signal to customers that they are ready to spend added resources to provide additional service at the prices they charge.

And fourth, prices distribute the costs of telephone service among users.

Costs

From a macroscopic view, the costs of producing telecommunications service consist of the resources that must be invested in the various components of the network plus the ongoing expenses of keeping it running. These components add up to the capacity to provide some maximum amount of service.

Greatly simplifying the details, let's consider a generic telephone call—or a quantum of usage of the network—and a single type of capacity. Up to some point, more telephone calls can be made, but then beyond that point service quality very quickly begins to deteriorate—some calls cannot be completed, some subscribers don't get immediate dial tone, and so on. There are definite limits to the capacity of the network.

The costs of a telephone network can be broadly classified. Some resources are simply fixed—they are required whether there is one customer or a million, whether the network is handling any traffic or not. Other resources are solely dedicated to serving an individual customer and are not needed if that customer is never connected to the network.

The largest proportion of the resources invested in the network is for capacity that is shared and used in varying degrees at different times by several customers. When one customer makes more calls, a neighbor may, or may not, be also able to make additional calls, depending on how much capacity is available. And the costs of providing this capacity depend almost entirely on the *maximum possible* rate of calling, not on how many calls are actually made.

Variable costs in the telecommunications industry are probably among the lowest of any industry in the economy. It takes fuel to move resources through a transportation network; in the electricity network, fossil fuel energy is first transformed and then transported to customers. In both cases, greater output requires more fuel or more energy. But in communications, the costs that vary in proportion to the number of calls are almost negligible.

» Achieving Efficiency in Local Telephone Pricing

In a market economy, prices are the primary instrument for allocating economic resources to their highest valued uses and promoting efficient production of goods and services. How well do local telephone rates perform this role? A recent RAND study (Park and Mitchell 1987) examined flat rates, measured service rates, and other possible time-of-day rates.

For the study, the local telephone market was reduced to a single measure of output—an average call of average duration. Calling rates were assumed to be constant during each hour and to vary by hour during the day. The "load curve," which plots number of calls by hour of the day, will vary with the particular features of each exchange. Typically, the greatest demand for calls occurs at only a few times during the day—late morning, mid-morning, or late afternoon.

Traffic engineers and investment planners must predict these peak demands in order to determine network capacity requirements. When the hourly demands of an entire year are plotted on a single graph, with the very peak load in hour number 1, the next highest load in hour 2, and so forth, the result is a waterslide-shaped figure called the "load duration curve."

The load duration curve reveals two key points. First, a very few hours of the year are responsible for a very large proportion of the maximum demand. And second, those peak hours are not bunched together in any natural order. The busiest hour might be at 11:00 a.m. in late November, the next busiest hour at 3:00 p.m. in the first week of July, and so on.

Faced with these demand patterns, telephone companies have generally not attempted to provide 100 percent service. Instead, they consciously plan capacity so that about 99 percent of the busy-hour calls are completed with a high quality of service, and the rest are lost or redialed.

Thus, the long-term planning process involves balancing the resources invested in network capacity, pricing access and use of the network, and managing network loads.

Load Management

In the telephone network, load or demand management most often involves simply not serving new calls that are offered to the network. Economics calls this form of load management "quantity rationing." It is a policy of assigning the available capacity randomly among potential users on a first-come, first-served basis. In special circumstances, such as a natural disaster or switching office fire, load choking procedures may also be used. In each case, these strategies are based on waiting for excess demand to materialize, and shedding load when it occurs.

Another load management strategy is to charge for the use of the network during hours in which demand is expected to be high. This is a strategy of "price rationing"—consciously attempting to modify the load curve in advance. Busy-hour usage is reduced by making it costly for subscribers to make calls at those times of day.

In practice, local measured service rates combine the strategies of price rationing and quantity rationing. Usage is priced at the highest rate in the hours of highest volume; in addition, capacity is limited and some calls are not completed during busy hours.

Although time-of-day rates can help to limit demands for scarce capacity, they are an imperfect tool for managing loads, for two reasons. First, the level of calling varies from one hour to the next within whatever fixed peak pricing period is set. Second, the actual loads that will occur in each hour are not really known in advance, at the time the rates and rate periods are established. As a result, when actual load is below the available capacity, the peak rate unnecessarily limits use of the network (Gale and Koenker, 1982; Vickrey, 1971).

» *Dynamic Rates: An Idea Whose Time is Arriving?*

The idea of dynamic rates has been advanced for several network services. It is a possible method of balancing loads, limiting congestion, and avoiding the high costs of adding capacity.

One proposal for the highway network envisages sensors embedded in the roadway that would record, probably magnetically, the license numbers of passing vehicles. Road usage at congested hours would be priced at an increasing rate, under control of a central computer, and drivers would be billed periodically. A crude version of this system is found in freeway on-ramp gates in some cities. Those gates limit the frequency of freeway access, imposing a waiting-time cost on drivers wishing to use the network.

The electricity network frequently supplies selected customers with power on an interruptible basis. The tariffs or contracts variously allow the utility to curtail supply on short notice, or to charge for consumption at premium rates, when systemwide demand approaches overall capacity.

In the telephone network, early proposals for managing peak loads included alerting subscribers to peak periods with a higher pitched dial tone and charging premium rates for calls made at those times (Vickrey 1981).

Telephone service is likely to be the very network service that is most readily priced in a dynamic form. The telephone network inherently possesses the capability for two-way communications to the customer, unlike the electric power and transportation networks. And although traditional telephone sets are still widely used, "intelligent" consumer terminals with some microprocessing features are spreading rapidly. One can readily imagine extending the capability of a telephone that can now store and recall frequently dialed numbers at the press of a button. It might, for example, also display the price of a telephone call, or it might simply activate a red light on a signal from the central office that peak pricing is in effect (Holm 1977).

How Might Dynamic Rates Work?

Dynamic telephone rates are still just an idea. But one possibility is that load management software in the central office would continuously monitor network performance. As loads build up close to peak capacity, the computer would set a price for network use. To do this, it would forecast the loads expected in the next few minutes, allow for demand elasticity based on recent experience, and set a price that would limit loads to a bit less than full capacity.

The price would remain fixed for a short period, say five minutes, and would be quoted automatically to calls in progress and to subscribers lifting the handset to place a call during the period. Consumer telephones would display the current price, or turn on a red light.

The central billing system would bill calls at these dynamic usage rates and provide summary statistics on consumers' monthly bills, with an indication of the maximum rate in effect for each call that occurred in peak periods. Over the longer term, some revenue from peak-period rates would offset investment in expanded network capacity.

At least initially, dynamic telephone rates would probably not be widely popular. The certainty of a known monthly bill appeals to many residential subscribers who might otherwise reduce their local telephone costs somewhat with measured service rates today. However, dynamic rates would offer price-elastic subscribers the opportunity of having telephone service for only the price of access service, provided that they avoid making calls when the peak-price red light is on.

A dynamic rate would almost surely be offered as an optional tariff, allowing off-peak and price-elastic users to reduce their telephone bills. If dynamic rates merely appealed to subscribers who already call only in nonpeak hours, they would be of no benefit to telephone companies seeking to reduce peak loads or recover costs of added capacity. However, several types of demands are likely to be price-sensitive. One-way messages, including some voice mail and facsimile, require no response and can often be postponed without loss. Information calls and data base inquiries that are answered by automatic equipment are readily deferred. And many casual conversations lack urgency and can occur at a later time.

Despite this potential, load-shifting may initially be low. But personal computers, electronic mail, and intelligent telephones will continue to develop rapidly and converge. Before long, we can expect consumers to have easy-to-use equipment that will send messages automatically at low rates, schedule data transmissions economically, and download central information when rates are lowest. And business subscribers with heavy data communications needs will find it economical to use off-peak intervals to transmit nonurgent bulk data.

Dynamic rates raise other issues that will need investigation. The telephone utility's revenues will be more volatile, subject to the chance occurrence of very high volume calling in a few hours a year.

New or extended network equipment and software will be required. And rate policy will need to be coordinated when several networks are used to complete a call.

The Future

Are dynamic rates a part of the local exchange of the future? Perhaps not soon. But the local network is expected to evolve and become even more capital-intensive as it is extended to provide switched broadband digital services. The gains from allocating capacity dynamically on demand will be larger, and the capabilities of consumer terminal equipment will have expanded markedly. In this environment, dynamic rates will enable telephone companies to achieve higher overall use of network capacity and will allow price-sensitive customers to shop for communications bargains on a daily basis.

» *References*

Gale, William A., and Roger Koenker. 1982. "Pricing Interactive Computer Services: A Rationale and Some Proposals for UNIX Implementation." Technical memorandum 82–11214–1, Bell Telephone Laboratories.

Holm, C.J. 1977. "The Responsive Telephone." Memorandum, British Post Office (16 May).

Park, Rolla Edward, and Bridger M. Mitchell. 1987. "Optimal Peak-Load Pricing for Local Telephone Calls." Publication R–3404–1–RC. Santa Monica, Calif.: RAND Corporation (March).

Vickrey, William. 1971. "Responsive Pricing of Public Utility Services." *Bell Journal of Economics and Management Science* 2, no. 1 (Spring): 337–346.

Vickrey, William. 1981. "Local Telephone Costs and the Design of Rate Structures: An Innovative View." Mimeo (March).

» PART V

Perspective of the Practitioners: The Cost Allocation Drivers

17 » Cost Allocation Techniques and Pricing Alternatives: Crossing the Great Divide

Michael L. Goetz

» Introduction

Cost allocation in telecommunications is so ubiquitous that to imagine what the industry would look like without it is virtually impossible. The views of accountants and economists on cost allocation are well known—and almost too predictable. Accountants view cost allocation simply as part of the necessary record-keeping that must go on within the firm and ascribe no normative content to it.[1] Economists consider most cost allocation unnecessary at best, and harmful at worst, because it is inherently arbitrary and diverts attention from marginal cost (Baumol, Koehn, and Willig 1987: 16–21). Although trained as an economist, the author's sentiments are with the accountants. Cost allocation has, and will continue to have, a prominent place in the telecommunications industry.

That cost allocation has a prominent place does not mean that its effects cannot be harmful. One of the principal arguments of this paper is that it is the combination of cost allocation rules with the current regulatory framework that causes many of the problems attributed by economists to the allocation of costs. This paper examines the linkages between cost allocation and pricing in a framework where both efficiency and fairness matter. Striking the proper balance between fairness and efficiency in the current regulatory environment is a most difficult task.

The first section asks the slightly heretical question: Why are costs allocated? The answers are found in the technology of telecommunications and in the requirement within a regulatory environment that there be a set of rules to follow. The next section briefly discusses the cost allocation literatures from economics, accounting, and game theory. To a large degree, these three disciplines reach consistent results, but employ different terminologies. A detailed example is constructed that allows us to move between the accounting and economics literatures.

The third section analyzes the costing and pricing options inherent in the Federal Communications Commission's (FCC's) price cap proposal and in the attempts to separate regulated from unregulated costs. Cost allocation rules have an important role to play in both of these policy issues. The last section attempts to offer something unusual for anything written by an economist: practical advice.

» *Why Are Costs Allocated?*

Simply put, costs are allocated in order to compute unit prices for firms that produce multiple products and whose technology is characterized by joint or common costs. If either of these conditions does not hold, the allocation of costs is not required. If the technology has no costs that are joint or common, production is nonjoint and it is possible to calculate unambiguously the costs of producing each of the firm's outputs. The sum of these costs will equal the firm's total cost; in economic terms, the cost function is separable or additive. If the firm makes only one product, cost allocation is simply unnecessary. Since both of these conditions do not hold for virtually all real firms, the allocation of costs in some form is ordinarily required.

The cost allocation problem is compounded in two additional ways, both of which serve to plant us firmly in the real world. First of all, cost allocation is required because the cost information produced by accounting systems in the presence of joint production and multiple products is necessarily arbitrary. These systems are attempting to divide the indivisible, and the results stand as cost accounting rules or standards. While these rules can be described as arbitrary, they can also be described as a proper response to a situation where precise answers are simply not possible and rules are employed to cope with complexity.[2] Second, while the rules represent a proper response, as with all rules they tend to become codified. Once codified, however, the rules are very difficult to change. The constituencies that benefit the most under the current regime will argue for the status quo, just as those who are relatively disadvantaged will fight for modification. Change will occur only when the combination of internal and external pressures are sufficient to overcome the inertia of the rules.

Costs are allocated first of all because of the technology of the firm or industry. The allocations are perpetuated because of informational limitations and inertia. The technological preconditions can only be altered slowly. The inherently arbitrary nature of cost allocation is not harmful in and of itself. It is the codification of the rules that poses the greatest difficulty, especially in a regulatory setting. Once determined, the cost allocation rules can be changed only after allowing all affected parties the chance to make their objections known. Such a process is protracted and time-consuming but allows all concerns, both reasonable and unreasonable, to be aired in a public forum.

Cost allocation in a partially regulated setting results in the worst of both the competitive and regulatory worlds. Quite simply, competitors do not care about the specifics of cost allocation debates. While regulated firms and commissions are discussing a particular allocation formula, competitors are engaged in calling on customers. The regulatory mechanism finds cost allocation very troublesome when a firm is only partially regulated. Charges of cross-subsidization will be made, the result of which is a long, protracted process of punch and counterpunch by the impacted parties. Both fairness and efficiency are sacrificed in this process.

The deliberate nature of the regulatory process is increasingly troublesome because technology and comptition are moving at an almost frantic pace. Indeed, this tension was clearly observable in the process of submitting cost manuals to the FCC, and in the discussion to date on price caps. Competition demands an immediate response, perhaps based on a guess; regulation, to be fair, requires a full and time-consuming hearing. While costs are allocated for entirely understandable reasons, the results are particularly harmful when these procedures coexist with a regulatory mechanism that of necessity can only accommodate gradual change. Under the current system, "fairness" is defined procedurally.

As it turns out, the academic literature on cost allocation can resolve *some* issues on which substantial energy is being spent. As with all answers that come from the academy, it cannot answer the questions confronting practitioners as specifically as would be desired. Partial answers, however, are still useful.

» *The Cost Allocation Literatures*

There are several distinct literatures on cost allocation, with varying degrees of attachment to the disciplines of accounting and economics. The paper will examine the accounting, game theoretic, and economics results on cost allocation methodologies. This section of the paper will summarize the results from these literatures with the goal of providing practical, concrete guidance to those who have the responsibility for responding to regulatory and company initiatives on cost allocation.

For the most part, the accounting literature on cost allocation is concerned with the degree to which cost allocation rules are based on verifiable and objective information. The search for objective and verifiable information leads rather easily to a set of rules or procedures that can be implemented using available information in an externally verifiable fashion. This view of cost allocation is consistent with the accountant's function of providing a historical record of transactions for consistency, internal and external, to the firm. "Verifiable and objective" is entirely consistent with "arbitrary and capricious." Cost allocation is often required for tax calculations and by transfer pricing mechanisms, and it does little good to lament its limitations.

Most of the accounting literature is positive in tone. That is, it is concerned with the consequences of various allocation mechanisms; it is not concerned at all with why costs are allocated. This is perhaps one of the reasons that accountants and economists have often disagreed so vigorously about cost allocation. Accountants view cost allocation as a normal part of the process of doing business and seek to examine the results. Economists seek to eliminate what

they believe to be unnecessary behavior by arguing that cost allocation can produce undesirable results, which indeed it can. One side
seeks to explain that which the other side argues should not exist.

Recently, game theory has joined the cost allocation debates.
The axiomatic (game theory) has produced a set of rules that
guarantee that (1) all costs are assigned to a product, and (2) no
more (or no less) than total costs, including a rate of return to capital,
is assigned to the list of products. These results currently apply to
only a restricted category of cost functions. Moreover, these axiomatic rules make substantial demands on the firm to generate rather
detailed and complex information. While the results are theoretically
precise, they are practically inapplicable.

The economics literature argues that cost allocation is arbitrary, and efficient only by accident. Cost allocation is viewed as
being antithetical to cost causation. From the economics cost
perspective, a cost function is intended to be both forward-looking
and a reflection of the resources employed in the production of
the firm's output. In this context, cost allocation smacks of arbitrariness in that certain costs are "assigned" or "allocated" to
various outputs independent of any causal linkage. This critique
is more compelling in the long run, where all costs are variable.
In the short run, however, the existence of discontinuities, capital
lumpiness, and nonvariable costs makes cost allocation necessary,
especially in a multiproduct environment. Even in the long run,
knowledge of the actual production process may require that some
costs be allocated.

Fortunately, there is an approach that retains the essentials of
the economist's paradigm (which allows for some precision) but links
it to the reality in which firms actually operate. The key concept
in this approach is the concept of economies of scope.

Economies of scope represent the cost savings attained by the
mix of products produced by the firm. That is, economies of scope
are said to exist if it is cheaper to produce a mix of outputs within
one firm rather than having a group of firms, each of which
specializes in the production of one of those products, produce the
same mix of outputs. In fact, if it were possible to divide up a firm
into a series of pieces with no overlapping costs, there would be no
advantage to multiproduct firms on the cost side. Note also that
in this hypothetical case, there would be no need for cost allocation, since all costs could be uniquely and unambiguously assigned
to individual products.

Economies of scope, then, result from inputs that are shared or used jointly in the production of several products. A university library can be used by both undergraduates and graduate students. These shared resources that create scope economies can also be called "indirect" costs, that is, costs that cannot be assigned to individual products. In other words, those resources that cause scope economies must of necessity be allocated to individual products, since by definition they are nonassignable.

The multiproduct cost model that is proposed meets several criteria. As presented, the model is consistent with the dictates of economic theory. This consistency is not sought for purposes of elegance but rather for both ease of interpretation and the subsequent ability to forecast. By employing the structure of economics, one can call upon a framework within which any results can be interpreted; this framework can also allow forecasts to be not only made but evaluated more easily. The basic form of the model is given by

$$C = f_o + V(Y) \tag{17.1}$$

where C represents a measure of total cost, the sum of fixed and variable cost (allocated expenses are included). f_o is a measure of fixed costs, $V(Y)$ is a measure of variable costs, and Y is a vector of outputs.

In words, equation (17.1) indicates that there is a functional relationship between total cost C and its division into fixed and variable components f_o and $V(Y)$. In this form of the model, variable costs depend on the output vector Y, and fixed costs are independent of the output vector Y. The assumption is made here that product-specific fixed costs cannot be properly defined, and therefore, fixed costs exist only in the aggregate.

This assumption of only aggregate fixed costs is quite extreme. The relation of this assumption leads to the second specification of the model, which is given by

$$C = f_o + \sum_{i=1}^{n} f_i + V(Y) \tag{17.2}$$

The form of (Y) is the same as in equation (17.1) as are the definitions of the principal variables. Fixed costs, however, are now equal to

$$f_o + \sum_{i=1}^{n} f_i. \tag{17.3}$$

The term f_o represents an overall fixed cost, that is, a cost that must be incurred to produce the output vector but is not functionally related to that vector. The term f_i represents the fixed costs that are specific to the production of output i. Incremental variable cost, IV_i, can be calculated from $V(Y)$.

Whether one is discussing equation (17.1) or equation (17.2), the form of the inclusion of fixed costs is extremely important in these models. First of all, fixed costs must be explicitly accounted for, since in a short-run cost model some inputs cannot be varied. Most important, fixed costs, either product-specific or overall, represent a key source of economies of scope—the cost advantages attributable to multioutput production.

The concept of economies of scope is crucial in the evaluation of the cost structure of any multioutput firm, since economies of scope represent the rationale for the multioutput firm. Fixed costs in and of themselves can generate economies of scope. Fixed costs of sufficient magnitude can even overcome the effects of interproduct cost penalties, or if cost complementarities exist, can serve to reinforce them. Therefore, the various cost models will be designed so as to incorporate these various cost measures.

The ability to move between the accounting and economic perspectives is summarized in Figure 17–1, where the accounting (direct/indirect) and economic (variable/fixed) classifications are related to one another. This relationship allows the analytic capabilities of the economic model to be linked to the strategic and operational language of the accounting analysis.

Given information on the aggregate cost of producing the range of services for a given month, Figure 17–1 allows this total cost of service to be divided up in several different ways. The direct cost of any service is equal to those costs that vary with the level of that service (column 1), plus any fixed costs (column 2) that are uniquely associated with that service. If all the direct costs associated with the set of products provided were added up and subtracted from total costs, the remainder would be indirect costs (3+4). These are the elements of costs that are shared among products or services and that cause the production of multiple services to be cost-effective relative to stand-alone production. By examining the columns, measures of variable cost (1 + 3) and fixed cost (2+4) for all services are obtained.

FIGURE 17–1

» *A Taxonomy of Costs.*

	Variable Cost	Fixed Cost
Direct Cost	1	2
Indirect Cost	3	4

The relationship in equation (17.4) must always hold, that is, the total of accounting measures of cost must equal the sum of the economic measures of cost. The conceptual scheme in Figure 17–1 allows an economic cost model to be estimated and the results to be subsequently translated into accounting terminology.

$$\text{Direct Costs} + \text{Indirect Costs} =$$
$$\text{Variable Costs} + \text{Fixed Costs} \qquad (17.4)$$

In Figure 17–2, the cells of Figure 17–1 are filled in, in terms of the cost function of equation (17.2). In Figure 17–3, various measures of cost are defined, consistent with the cost function of equation (17.2). That is, an economic cost function is constructed that explicitly links the accounting and economic conceptions of cost. Costs are allocated, but the allocations are based on an economic framework. Economies of scope provide the glue that links the approaches together.

FIGURE 17-2

» *A Cost Equation of the Form.*

$$C = f_o + \sum_{i=1}^{n} f_i + V(Y)$$

and

$$IV_i = \Delta V/\Delta Y_i$$

where C is the total cost
 f_o is a fixed cost for all products
 f_i is a fixed cost for product i
 IV_i is the incremental variable cost for i
 $V(Y)$ is a variable cost for all products

	Variable Cost	Fixed Cost
Direct Cost	$\sum_{i=1}^{n} IV_i$	$\sum_{i=1}^{n} f_i$
Indirect Cost	$V(Y) \sum_{i=1}^{n} IV_i$	f_o

FIGURE 17-3

» *Various Cost Measures (derived from Figure 17-2).*

For Any Product:

Direct Cost $= IV_i + f_i$

Marginal Variable Cost $= \Delta V(Y)/\Delta Y_i$

Fixed Cost $= f_o + \sum_{i=1}^{n} f_i$

For the Set of Products:

Economies of Scope $= \dfrac{f_o + [V(Y) - \sum_{i=1}^{n} IV_i]}{C} = Sc$

Economies of Scale $= \dfrac{C}{\sum_{i=1}^{n} \Delta V/\Delta Y_i \, Y_i} = S$

» *Two Examples*

The title of this paper includes mention of a great divide. There are two interpretations of the term "great divide." It can refer to the process of cost allocation itself, or to the gulf created jointly by economists and accountants on the necessity of allocating cost. This paper has taken the position that there is indeed hope if it is recognized that not all issues raised in the cost allocation debate can be answered. Some can be answered, however, in such a way as to provide useful policy advice. Two current policy issues will be briefly discussed in terms of the composite cost allocation model.

Price cap deregulation proposals have been put forward at the federal level and in some states. These proposals all build on the British approach to telecommunications deregulation. As proposed by the FCC, implementation of price cap–style regulation requires measures of cost by product, a mechanism to adjust the cost standards for changes in productivity, and agreement on the definition and degree of disaggregation of services. Measures of product-specific cost and the attendant product definition presuppose some type of cost allocation. To assume that cost need not be allocated is to deny the reality of telecommunications as the industry currently exists. Costs can, however, be allocated in an economically consistent, albeit somewhat arbitrary, way.

As put forward, the FCC's price cap proposal (FCC 1988) employs cost measures with precise economic definitions. Indeed, the references to the appropriate literatures attest to the intentions of the authors. The important point to note is that these economic cost measures (for example, stand-alone cost) are not at all related to their separated cost counterparts. Stand-alone cost is the cost associated with producing a single output at its current level, *with all other outputs zero*. To implement price caps requires that cost measures be developed that are externally verifiable; this will exclude accounting-based cost allocations.

Similar problems arise with the definition of product groupings. These product categories should be defined in a cost-causative manner. That is, product groups should be defined such that the four categories in Figures 17–1 or 17–2 can be filled in. This classification scheme defines direct and indirect costs by product group. Direct costs as defined are the appropriate floor for cross-subsidy tests. Indirect costs must be assigned to the product groups such that, in the aggregate, they are just covered.

If implemented, price caps will exhibit the following tendencies. Cost allocation will still be required and can be expected to become even more controversial. Price caps will be no more flexible than current regulatory standards. This prediction has less to do with the proposal itelf than with the remnants of regulatory procedures that are retained under the FCC proposal. Ultimately, the information requirements associated with price caps will create an analog to the existing process unless there is agreement on the empirical implementation of price caps.

The second policy issue involves the separation of the costs of regulated activities from those of unregulated activities. The joint cost order still mandates the use of a fully distributed cost (FDC) methodology. Concerns for cross-subsidization are indeed appropriate and should be given primacy of place in policy debates. An FDC methodology, however thoughtfully designed, can never effectively address the magnitude or direction of cross-subsidy because the methodology implicitly contains the very cross-subsidies it is designed to detect. The advantages of integration, or economies of scope, are not adequately reflected in the order. Emphasis should be placed on the most accurate measure of direct cost by product and on the aggregate of indirect costs as measured by economies of scope. The arbitrariness of the procedure should be faced directly and be openly acknowledged. A disservice is imposed *on all parties* if one continues to pretend that costs cannot and should not be allocated. There is a proper way to proceed that requires that the reasons for cost allocation be made clear and that the accounting and economic approaches to that allocation be made as consistent as possible.

To conclude, the paper has tried to argue for cost allocation when it is defined within an integrated economic-accounting framework. These arguments can be summarized in the following four propositions:

1. Cost allocation will continue to play a significant role in telecommunications.
2. While arbitrary to a degree, cost allocation rules need not be capricious or harmful.
3. Balancing the requirements of fairness, business prudence, and competition will require that the outcomes of the cost allocation process become more important than the process itself; that is, the precise nature of the cost allocation criteria is much *less* important than the prices that result from the implied allocation.
4. The best set of cost allocators must balance flexibility and simplicity and be compatible with competition.

» *References*

Baumol, William J.; Michael F. Koehn; and Robert D. Willig. 1987. "How Arbitrary Is Arbitrary? or, Toward the Desired Demise of Full Cost Allocation." *Public Utilities Fortnightly* 120, no. 5 (3 September): 16–21.

Brown, Stephen J., and David S. Sibley. 1986. *The Theory of Public Utility Pricing.* New York: Cambridge University Press.

Cohen, Michael D., and Robert Axelrod. 1984. "Coping with Complexity: The Adoptive Value of Changing Utility." *American Economic Review* 74, no. 1 (March): 30–42.

Federal Communications Commission. 1988. "Further Notice of Proposed Rulemaking in the Matter of Policy and Rates Concerning Rates for Dominant Carriers." CC Docket No. 87–313, FCC 88–172 (adopted 23 May 1988).

» *Notes*

1. For a good summary, see Brown and Sibley (1986).
2. This view is based on an approach in which transactions costs are treated in a nontrivial way. For a model with this flavor, see Cohen and Axelrod (1984: 30–42).

18

» *Costing and Pricing in the Telecommunications Industry: The Formula Is Everything*

Anthony G. Oettinger

» *Reckoning Costs and Devising Prices*

The formulas were the key. That is to say, the ways under which the benefits of various Federal programs are calculated. Invariably, these formulas favor some sections of the country over other sections . . . Senator Hill of Alabama was a kind and courtly man, properly concerned about his region of the country. Hence the formula for Hill-Burton funds [for hospital construction]. Programs still going compensate low-income states in proportion to the square of their distance from the national average. In 1977, in a commencement address at the Kingsborough Community College, I suggested the square root. The formula is everything.

> —Senator Daniel Partrick Moynihan
> *Letter to New York*, 19 September 1987

This paper deals with the challenge of inventing ways of reckoning costs and devising prices—ways that are in keeping with the consensus of the times.[1] From a perspective that views the definitions of products, services, their costs, and their prices as discretionary, such questions as, What are the true costs? and, What are the associated cost-based prices? amount to hunting the unicorn.

Consider, instead, costing and pricing as tools. What costing objectives? To convey what messages to suppliers and to customers? What costing methods and what pricing methods lend themselves well to practical administration? Rather than dealing in absolutes— unless you're blindly committed to your own beliefs as unique and infallible, or committed by professional duty to a cause or a party line—consider a more productive way instead. Consider cost and price definitions as fairy tales whose merit lies in how well they meet the needs or the goals of various stakeholders—companies, customers, regulators, politicians, and so on.

Whatever contending theologies or party lines are in vogue, in practice the prevailing costing and pricing methods reflect more or less faithfully, and with greater or lesser time lags, the prevailing political balances of their day. At their best, costing and pricing methods are the means to policy ends, not ends in themselves.

What policy ends? Those in harmony with whatever consensus or compromise is acceptable to the stakeholders and to the referees involved in the battle: providers, customers, competitors, regulators, legislators, the courts.

Drawing on the past as well as on the present for its foundations, this paper is meant to help us all cope with the protracted uncertainties that accompany continuing guerrilla warfare—a guerrilla warfare not only within the telecommunications industry but also at the borders between telecommunications and other information industries, such as the computer and newspaper industries.[2]

Starting with the big picture and working down into concrete and specific detail, this paper will highlight, at each step, where substantial discretion was or could have been exercised.

In order to give sharp illustrations of what remains an inherently fuzzy situation, data will be referred to from work in progress at the Harvard Program on Information Resources Policy.[3]

» Discretion in Cost/Price Relationships

Milestones of Telecommunications Cost/Price Relationships

Folks love to spread the idea that the prices of products or services are related to the costs of those goods, but it ain't necessarily so.

In actuality, prices sometimes have little to do with costs. Sometimes prices and costs are tightly bound together. Mostly, however, the relationship between prices and costs swings back and forth, like a pendulum in a clock, between one extreme and another.

The degree of linkage between prices and costs depends more on the political consensus of the moment than on some absolute reality. That consensus is called the "prevailing philosophy," or the "accepted policy," by those who like it. That same consensus is "dirty politics" to those who don't like it. Either way, policy objectives and administrative tools are often not synchronized.

For telecommunications, Figure 18–1 highlights some of the major philosophical/political shifts of the last half century. The United States is in the midst of another major shift right now.

The remainder of this paper illustrates how each of these shifts did or did not bring costing or pricing means into harmony with the policy ends of its time. A picture emerges in which outcomes reflected political consensus more than wishful thinking, special interest, or professional dogma.

To put the cost-price relationship story in focus, a snapshot of one of the many issues in the telecommunications industry is used. The name of this snapshot is "Exchange Access Charges: 1986."

Focusing on exchange access charges blazes a trail through the events sketched in Figure 18–1. It helps us see how these events defined and redefined exchange costing and pricing methods—and telecommunications costing and pricing in general—according to the politics of the day.

First, the very definition of "exchange service" highlights how much discretion there is in that definition. Then there is further discretion in the definition of "access," and yet another layer of discretion in setting its price.

The Scope of Discretion: What You See Depends on Who You Are and When You See It

"When *I* use a word," Humpty Dumpty said, in rather a scornful tone, "it means just what I choose it to mean—neither more nor less."
—Lewis Carroll, *Through the Looking-Glass*

What Is Exchange Service?   Access charges are the price paid to so-called local operating companies, or local exchange carriers (LECs) for what the Modification of Final Judgment (MFJ) calls

FIGURE 18–1

» *Milestones in Cost/Price Relationships, 1930–1987.*

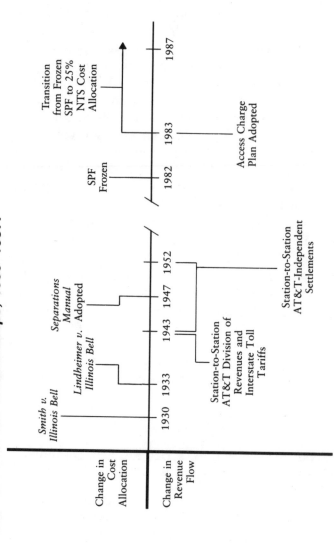

access services, namely, "the provision of exchange services for the purpose of originating or terminating interexchange telecommunications" (MFJ 1982: sect. IV[F]).[4]

Every object in the MFJ's definition, including exchange service itself, is discretionary, not an immutable natural object or an immutable law of nature. The discretion may be individual or social, immediate or realizable only over generations, but it is discretion nonetheless.

Much discretion runs counter to the commonly held notion that a price is the price of something definite and that a cost is the cost of something definite. Mostly, however, that something is not at all definite, but in constant flux. The thing being costed or priced often will differ tomorrow from what it is today. The thing being costed or priced often differs today from what it was yesterday.

What, for instance, is local or exchange service for which there may be access charges? The Communications Act of 1934 defines it flexibly, even circularly:

> "Telephone exchange service" means service within a telephone exchange, or within a connected system of telephone exchanges within the same exchange area operated to furnish to subscribers intercommunicating service of the character ordinarily furnished by a single exchange, and which is covered by the exchange service charge.

Exchange service, in plain english, is that which is charged for as exchange service. The letter of the law has stayed put since 1934, but what is charged for keeps changing with passing time, as in the growth of extended area service (EAS), which expands the geographic scope of exchange service and reduces the scope of interexchange service. The juxtaposition of "local" with "exchange," as in "local exchange carrier," is therefore, at best, an anachronism. For instance, the metropolitan Atlanta area "covered," in the words of the Communications Act of 1934, "by the exchange service charge" is about the same size as the whole state of Rhode Island.

The precise location of the exchange/interexchange boundary might or might not make a difference in the concepts of access, the cost of access, or the price of access. For example, without changing either cost or price one penny, one can hold that access is either a feature of interexchange service or a part of exchange service. The official view sees access as interexchange, but most customers view the subscriber line charge (SLC) as part of what they pay for exchange service.[5] There are many layers of discretion.

The concept of access resembles a man-made fairy tale more than it resembles a mathematical relationship like the number pi. There is a lot of maneuvering room within which mortal entrepreneurs, mortal judges, and mortal commissioners can define access to mean whatever in their eyes suits the needs of the day.[6]

Discretionary Distinquishing Features of Products and Services. But even accepting some generic notion of access into and out of some accepted notion of exchange, there remains further discretion in defining precisely what access is, what access costs, and what price to charge for access. And stakeholders disagree over how to exercise that discretion.

One product or service is distinguished from another by its features, its cost to its supplier, its price to its buyer, and other characteristics. Preferences about features, costs, and prices usually vary among different stakeholders.

The distinctions are giving way among technologies, as more devices and systems go digital. But much of the world's technology is embedded in analog voice telecommunications, digital data processing, and analog broadcasting lines. Careers hang on preserving these boundaries, so the distinctions may remain, even without a technical difference remaining.

Although the boundaries between the voice and data industries are showing signs of crumbling here and there (IBM owns ROLM and part of MCI; AT&T sells computers), there's nary a crack in the wall between broadcasters and the computer and telecommunications industries.

The firm wall around the broadcasting business reflects the state of the law—and *reinforces* the state of the law. Title III of the Communications Act of 1934 separates over-the-air and cablecasters from the communications people of Title II of the act. Cynics see this as no accident, given that politicians see their re-election to office as more directly dependent on available air time than on the prices of telephones or personal computers.

Title II holds the telecommunications industry firmly in its grip, but it reaches the computer and other electronics industries with only an "ancillary to" hand. Hence the bright light of Computer Inquiry II never really shone. Instead, you have the foggy concept of open network architecture (ONA), as computer inquiries march from I through II to III and on toward infinity.

As of the late 1980s, the mass of *customers,* who know what they like when they see it, stuck to tradition. In most companies, the management information systems (MIS) manager and the communications manager were still worlds apart. The voice communications manager, who knew telephone companies and belonged to the International Communications Association, was in one world. The data communications manager, who knew local area networks (LANs) and belonged to a computer users' association like IBM's Guide, was in another world.

With all these differing perceptions, it is not at all surprising that there is disagreement over what products and services are (never mind what their costs to suppliers and their prices to customers are). The economists' rock-solid conception of products or services does not even apply to the real world.

One has only to think of the numerous historical shifts in the size and the geographical overlap of local service areas, or of the mid-1980s shift of responsibility for maintaining in-premises telephone wiring from the supplier to the customer. Since divestiture in 1984, AT&T and the Bell operating companies (BOCs) even disagreed over what it means to *make* a product. Does manufacturing include research and development, design, and so forth? Or is manufacturing only the building and aggregating of physical parts? In December 1987, the U.S. District Court for the District of Columbia ruled that it meant to include design in manufacturing—but the border wars continued.

What follows, however, ignores the plasticity of products and services themselves and assumes that we really *know* what a product or service is when we talk about its cost or its price.

» Outcomes of Discretion: A Snapshot of Access Charges 1986

This section sketches the size and significance of the access charge piece of the stakes in 1986. The next section then describes how discretionary answers, changing over the years in response to changes in the burning questions of the day, eventually mutated into the access charges of 1986.

As mentioned before, the concept of access charges occupies but one position in the swing of the pricing pendulum. Other stakes, such as the nature and extent of government intervention, swing on their own pendulums—sometimes in step, sometimes not. But the focus remains on the example of exchange access charges. The following discussion considers access charges in the context of total operating revenues.

What 1986 Access Revenues Amounted to

In 1986, the LECs of the United States charged others approximately $26 billion for access to their exchange services. Subject to some incompleteness in the data, and subject to some corporate fuzziness in reckoning the amount, that much is hard fact. This hard fact, however, rests on a discretionary foundation that was shaped, and continues to be shaped, by a shifting consensus among the stakeholders.

Of the $26 billion that LECs received in access charges, about $22 billion was paid to LECs by three interexchange carriers (ICs): AT&T, MCI, and US Sprint. The remaining $4 billion was paid to the LECs by either smaller ICs or their noncarrier customers. These payments were for either SLCs (federal or state mandated) or other access charges.[7] The breakdown of the $4 billion and its relationship to the total is an artifact of a possibly arrested transitional process outlined in Figure 18–2.

Somewhat fuzzier—but still hard enough—is the fact that the 1986 operating revenues of the LECs amounted to $80 to $90 billion. Included in this amount are revenues from unregulated ventures and from activities such as directory advertising and sales. These revenues amount to some $10 to $12 billion.

What one sees as the *telecommunications* revenues of LECs (or their parents) depends on whether revenues from items such as directory advertising are counted in or out. This is a matter on which practices differ among LECs.[8] It is subject to continuing review by the U.S. District Court for the District of Columbia, which, in 1987, retained and actively exercised authority over the structure of the divested pieces of the former Bell System.

What Customers Paid in 1986 to ICs and LECs

Figures 18–2 and 18–3 put the $80 to $90 billion total LEC operating revenues in the context of the pieces of the traditional industry.[9]

Figure 18–2 shows the aggregate of what end-user customers in 1986 paid to the three largest ICs, and what end-user customers (and

FIGURE 18-2

» *1986 Operating Revenues as Paid by End-User Customers.*

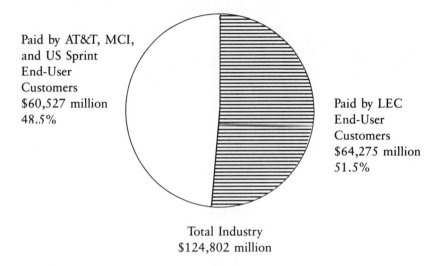

Paid by AT&T, MCI, and US Sprint End-User Customers $60,527 million 48.5%

Paid by LEC End-User Customers $64,275 million 51.5%

Total Industry
$124,802 million

Viewpoint: LEC end-user customer payments to LECs
IC end-user customer payments to IXCs

© 1987 President and Fellows of Harvard College. Program on Information Resources Policy.

some others) in 1986 paid to LECs. In aggregate, the customers who buy interexchange service between local access and transport areas (LATAs)—both interstate and within states—generate some 50 percent of the total revenues. The customers who buy exchange service (and interexchange service within LATAs) pay the other half of the total revenues. Some of what end-user customers pay to the major ICs is then paid by those carriers to the LECs. This brings the LEC revenues from the level in Figure 18-2 to the $80 to $90 billion level depicted in Figure 18-3.

With all the changes in industry structure and in the definition of service that happened in the 1970s, there is a remarkable stability in the overall proportions of Figure 18-2. From 1976 to 1980, roughly 50 percent of the total bill was paid by the buyers of what was then still mostly called toll service (interexchange service, both interstate and intrastate), and the other 50 percent of the total bill was paid by the buyers of exchange service.[10]

FIGURE 18–3

» *1986 Operating Revenues as Kept by Companies.*

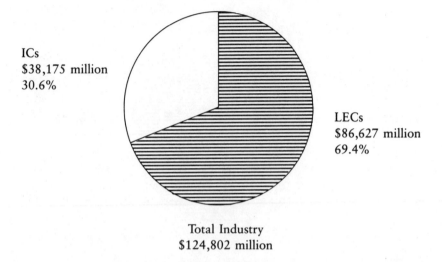

ICs
$38,175 million
30.6%

LECs
$86,627 million
69.4%

Total Industry
$124,802 million

Viewpoint: LECs as receivers of their end-user customer payments plus IC and
other access customer payments.
ICs after access payments to LECs.

The LEC revenues of Figure 18–2 do include some revenues
from intraLATA interexchange service alongside the exchange service
revenues. But how many local service areas make up a LATA—hence,
the proportions of exchange and interexchange communication—is
as discretionary a decision within the LATAs as it was within whole
states before LATAs were defined in response to the 1982 MFJ. Essentially, in the late eighties as in the late seventies, the proportion of
total payments for calls between down home and way out there and
for calls between down home and down home held in the 50–50
neighborhood.

The next section suggests how the lingering of a waning consensus supported this stability.

Who Kept Which Access Revenues in 1986

Figure 18–3 shows the aggregate of what the three largest ICs and the LECs kept for themselves. Approximately 30 percent of the total revenues are kept by the ICs, and about 70 percent are kept by the LECs.

The 20 percent difference between the numbers in Figure 18–2 and Figure 18–3 is made up of the payments that the three largest ICs make to the LECs for access to the LECs' exchange services. The aggregate amount was about $22 billion in 1986. In effect, the users of interexchange service pay for this access charge (levied by the LECs) as part of the price charged by the ICs for interexchange service. This is mostly but not entirely for service between exchanges in different LATAs.[11]

With the $4 billion or so in other access payments to the LECs by customers other than AT&T, MCI, and US Sprint, the total access charge revenue of the LECs totalled about $26 billion in 1986. The amounts and the proportions of these revenues reflect the 1986 balance between the politics of universal service and the politics of cost-causation.

In a rough-and-ready way, then, the $80 to $90 billion consists of revenues from more or less traditional telephone services rendered by the LECs with their intraLATA networks. This includes local access for both inbound and outbound interLATA traffic, as well as local access for intraLATA interexchange traffic handled by carriers other than the local LEC itself. It mainly includes the LECs' own exchange services and intraLATA interexchange services, a pricing distinction based on differences whose technological significance at the end of the 1980s continued to diminish toward the vanishing point.

Figure 18–4 shows how the total LEC revenues are distributed between the BOCs and the independent companies.

Table 18–1 details the access revenues and the operating revenues company by company. The high and low figures roughly bracket the uncertainty.[12]

Sensitivities to Changes in Access Revenues

Although the independents as a group get less total revenue from access charges than do the BOCs, some of the independents are more dependent than the BOCs on access charges. They are therefore more sensitive than the BOCs to changes in the level of access charge revenues

·FIGURE 18–4

» *IC, BOC, and Independent Shares of 1986 Kept Revenues.*

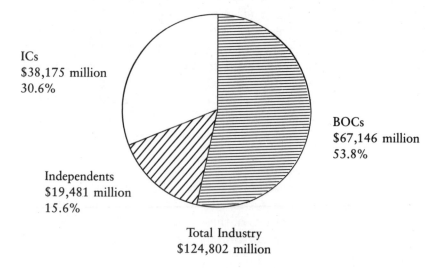

ICs
$38,175 million
30.6%

BOCs
$67,146 million
53.8%

Independents
$19,481 million
15.6%

Total Industry
$124,802 million

Viewpoint: LECs as receivers of their end-user customer payments plus IC and
other access customer payments.
ICs after access payments to LECs.

© 1987 President and Fellows of Harvard College. Program on Information Resources Policy.

and more vulnerable than the BOCs to changes in that level. This
phenomenon reflects historical cost patterns and a diminishing but
still lingering rural political power.

Sensitivities in the Large.   For the LECs as a whole, revenues from
access charges in 1986 amounted to about one-third of their operating revenues (see Table 18–2). But there is no such thing as an
average LEC, any more than there is an average person.

   At one extreme, revenues from access charges amounted to only
22 percent of the total operating revenues of Southern New England
Telephone (SNET). At the other extreme, the thousand or so very small
independent companies in the aggregate depended on access charges
for 69 percent of their revenues. Clustered around the average are the
BOCs held since their 1984 divestiture under the umbrella of the seven
Bell regional holding companies) and the larger independents.

» **Table 18–1.   1986 Local Exchange Company Access and Operating Revenues.**

	Access Revenues (\$ Millions)	*Operating Revenues (\$ Millions)*	
		Low	*High*
Ameritech	\$2,487	\$8,077	\$9,326
Bell Atlantic	2,685	8,224	9,921
Bell South	3,452	9,735	11,444
NYNEX	2,984	9,897	11,342
Pacific Telesis	2,695	7,540	8,867
Southwestern Bell	2,797	6,838	7,902
US West	2,595	7,570	8,308
TOTAL BOCs	\$19,695	\$57,881	\$67,110
Centel	397	833	1,370
Cincinnati Bell	104	430	492
Contel	982	1,835	3,074
GTE	3,357	11,277	15,112
SNET	317	1,433	1,433
United Telecom	924	2,372	2,372
Other Independents	698	849	1,237
TOTAL Independents	\$6,779	\$19,029	\$25,090
TOTAL LECs	\$26,474	\$76,910	\$92,200

© 1988 President and Fellows of Harvard College. Program on Information Resources Policy.

In 1986, then, the total income of all LECs depended fairly heavily on the aggregate of access charges. The collection of a substantial portion of these access charges was then still enforced by federal authority over a portion of telecommunications supplier costs.[13] In 1986, this portion of costs was still jurisdictionally separated in accordance with processes designed to support an earlier consensus that still commanded political respect.

Sensitivities in the Small.   Table 18–2 also highlights the LEC widely spread sensitivity to small changes in the level of the access charge revenue stream.

At one extreme, for the small independents as a group, a shift of one percent of total operating revenues from access charges to

» Table 18-2.   *1986 Local Exchange Company Sensitivity to Changes in Access Revenues.*

	Access Percent of Operating Revenues	Operating Revenues ($ Millions)	Effect on Access and Other Operating Revenues of Shifting 1% of Total Revenues from Access to Other		
			Percent Change in Access Revenue Level	*Percent Offset in Other Operating Revenue Level*	*Magnification (>1) or Reduction (<1) of Swing*
Other Independents	69%	$1,043	1.4%	3.3%	2.3
Contel	43	2,455	2.3	1.7	0.7
United Telecom	39	2,372	2.6	1.6	0.6
Centel	38	1,102	2.6	1.6	0.6
Southwestern Bell	38	7,370	2.6	1.6	0.6
Pacific Telesis	33	8,204	3.0	1.5	0.5
Bell South	33	10,590	3.0	1.5	0.5
US West	33	7,939	3.1	1.5	0.5
TOTAL BOCs	32	62,495	3.2	1.5	0.5
TOTAL LECs	32	84,555	3.2	1.5	0.5
TOTAL INDEPENDENTS	31	22,060	3.2	1.5	0.5
Bell Atlantic	30	9,073	3.3	1.4	0.4
Ameritech	29	8,702	3.5	1.4	0.4
NYNEX	28	10,619	3.5	1.4	0.4
GTE	26	13,195	3.8	1.4	0.4
Cincinnati Bell	23	461	4.4	1.3	0.3
SNET	22	1,433	4.5	1.3	0.3

other operating revenues would amount to a reduction of 1.4 percent in access revenues. To compensate for this shift, these companies would need to increase other operating revenues by 3.3 percent. The effect is a larger recovery from other services (read, greater increase in prices). The shift from access is magnified by 2.3.

Factors that might shift revenues from access to other services include changes in jurisdictional separations and a repricing of intrastate services that would reduce intrastate inter- and intraLATA access prices or interexchange prices and increase exchange prices. Such changes might be triggered by increases in bypass, changed definitions of what constitutes access, and other regulatory changes.

At the other extreme, the effect would be the opposite. The shift of one percent of total operating revenue amounts to shifting 4.5 percent of access revenues for SNET, with only a 1.3 percent change in other operating revenues required to offset this shift. Here, the offsetting shift is less than one-third of the shift in access charge revenues.

Figure 18–5 graphically portrays the relationship between how many dollars a company collects in access charges and how sensitive it might be to changes in that amount.

In the left-hand column of Figure 18–5, the shading gets darker as total access revenues decrease. In the right-hand column, companies are ranked by their sensitivity to changes in access revenues. However, these companies retain their shading from the left-hand column. For instance, the shading of "Other Independents" in the right-hand column indicates that although they are the most sensitive to changes in access revenues, the absolute amount of those revenues is among the least. At this stage, one can only speculate as to what accounts for the patterns in Figure 18–5.

Sources and Distribution of 1986 Access Revenues

Table 18–3 shows the sources and the distribution of revenues generated by access charges. Note that the numerous and geographically scattered small independents received only 2.6 percent of the 1986 LEC access revenues.

Also of continuing political significance for exchange pricing are the relationships between LEC access charges to the major ICs and other factors. One of these factors is the total of LEC access revenues, which includes revenues from SLCs. Another is the level of total IC charges to their customers.

FIGURE 18–5

» *Relation Between Sensitivity to Access Charge and Total Access Revenues.*

LEC Access Revenues ($ millions)	Rank		Access as Percent of Total Operating Revenues
Bell South	1		Other Independents
GTE Telephone	2		Contel Telephone
NYNEX	3		Centel Telephone
Southwestern Bell	4		United Telephone
Pacific Telesis	5		Southwestern Bell
Bell Atlantic	6		US West
US West	7		Pacific Telesis
Ameritech	8		Bell South
Contel Telephone	9		GTE Telephone
United Telephone	10		Bell Atlantic
Other Independents	11		Ameritech
Centel Telephone	12		NYNEX
SNET	13		SNET
Cincinnati Bell	14		Cincinnati Bell

© 1988 President and Fellows of Harvard College. Program on Information Resources Policy.

The high degree of dependence by the LECs on access revenues from the three largest ICs, especially AT&T, sets up a major customer–major competitor relationship that manifests itself in continuing political tensions between AT&T and the LECs. The high proportion of their total revenues that ICs spend on access charges is an incentive for the ICs to bypass the LECs and to push for government intervention to prevent others, such as end-users, from engaging in bypass.

Table 18–3 shows the access charges collected by LECs. These access charge payments of $22 billion amounted to over 50 percent

» **Table 18–3. 1986 Local Exchange Company Access Revenues as a Percentage of Total Industry Access Revenues.**

	Percent of Total Industry	Access Revenues ($ Millions)
PAYERS		
Access charges to AT&T, MCI, and US Sprint	84.4%	$22,351
Subscriber line and other access charges	15.6%	$4,122
TOTAL	100.0%	$26,473
RECEIVERS		
Bell South	13.0%	$3,452
GTE	12.7	3,357
NYNEX	11.3	2,984
Southwestern Bell	10.6	2,797
Pacific Telesis	10.2	2,695
Bell Atlantic	10.1	2,685
US West	9.8	2,595
Ameritech	9.4	2,487
Contel	3.7	982
United Telecom	3.5	924
Other Independents	2.6	698
Centel	1.5	397
SNET	1.2	317
Cincinnati Bell	0.4	104
BOCs	74.4%	$19,694
INDEPENDENTS	25.6%	$6,779
TOTAL LECs	100.0%	$26,474

© 1988 President and Fellows of Harvard College. Program on Information Resources Policy.

of the $42 billion of revenues that AT&T, MCI, and US Sprint collected from their customers—as revenues mostly, but not exclusively for interLATA interexchange services.

Table 18–3 also shows that no more than $4 billion, or 16 percent, of LEC access revenues in 1986 came from the phase-in of access charges paid by end-users—that is, the SLCs—and from other charges not flowed through AT&T, MCI, and US Sprint.

From work still in progress,[14] the best estimate of the Program on Information Resources Policy is that approximately $3 billion, or some 12 percent, of LEC access revenues in 1986 came from the SLC. The remaining $1 billion, or 4 percent, of LEC access revenues came from a mix of (1) ICs other than the three largest, and (2) a variety of end-users and intermediate users of special access and other access services.

Hence, 84 to 88 percent of LEC access revenues comes from payments to LECs by ICs. About 75 percent of those payments by ICs are for interstate access according to federal policy.[15] The remainder of the payments by ICs is for intrastate access according to the policies of fifty states and the District of Columbia.

LEC access revenues from ICs include payments by ICs for the carrier common line charges (CCLC) for access plant said to have non–traffic-sensitive costs, including the local loops, customer-premise equipment (CPE), embedded inside wire, and station equipment.

Notable among the many disjunctions between costing and pricing philosophies is the fact that much pricing for access is tied to traffic volume, although the costs said to underlie the prices are also said to be insensitive to traffic volume. These apparent logical disjunctions, common in the stream depicted in Figure 18–1, are based on a logic that is sketched in the next section ("Exercising Discretion"). They are as important in understanding the ways of international telecommunications as they are in understanding U.S. domestic policies and practices.[16]

The rest of the payments by ICs include charges for so-called traffic-sensitive costs (including switching, intercepts, information, and transport) and special access charges.

Although the bulk of access charges paid by ICs are for interstate and intrastate interLATA access, "interexchange" and "interLATA" are not synonymous.[17] Some intrastate ICs may also pay intraLATA access. Independent companies may buy access from LECs (and vice versa) in states where pooling has been replaced by an access arrangement. In the "corridors" around New York City and Philadelphia, and where there is interstate, intraLATA, interBOC interexchange service, as in Illinois east of St. Louis, BOCs may buy access from other BOCs. Still others, such as value-added networks (VANs) like Tymnet, Telenet, and so on, also buy access.

Moreover, not all buyers of access are carriers. For example, some large, sophisticated business customers may buy access, especially

FIGURE 18-6

» *1986 Largest IC Access Payments.*

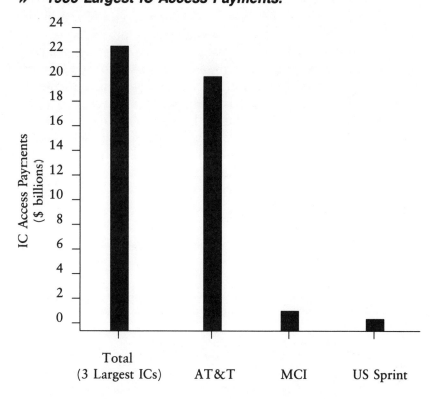

© 1988 President and Fellows of Harvard College. Program on Information Resources Policy.

if they have their own physical network. Further wrinkles include variations in the disposition of access revenues. By the end of 1987, almost all states let the LECs in their state keep the intrastate interLATA access charges that they billed. Only North Carolina clung to the formerly prevalent practice of state pooling of these revenues. But practices differed more for intraLATA toll revenues. Several other states were with North Carolina in continuing pooling. Florida, Kansas, and Missouri switched to "bill-and-keep," at the start of 1988. Bill-and-keep was already in place in most states.

In 1986 these exotic-seeming variations accounted for, at most, $1 billion, or 4 percent, of access revenues—and more likely, for less than half that amount. To some, however, they stand for the wave of the future. Hence, they account for more controversy than their small relative size at the end of the 1980s would suggest.

Figure 18–6 focuses on where most of the access money actually came from in 1986, namely, the three largest ICs. A glance back at Figures 18–2 and 18–3 then gives a graphic sense of what portion of prices paid by the end-users of interexchange services flows through the ICs to pay for costs of plant belonging to the LECs and for related expenses. And about half of the payments to the ICs by customers flow on to the LECs as payments for access.

There can be, and indeed there have been and continue to be, numerous controversies over how to look at the costs of "local" plant. How these costs are viewed may or may not have an impact on what prices end-users pay. Shifts in how stakeholders view costs of local plant is the main topic of the next section.

» *Exercising Discretion: Access Charges by Other Names*

> "Would you tell me, please, which way I ought to go from here?"
>   "That depends a good deal on where you want to get to," said the Cat.
>   "I don't much care where—" said Alice.
>   "Then it doesn't matter which way you go," said the Cat.
>   "—so long as I get *somewhere*," Alice added as an explanation.
>   "Oh, you're sure to do that," said the Cat, "if you only walk long enough."
>
> —Lewis Carroll, *Alice in Wonderland*

With the aim of discerning some key forces and trends, this section reviews some of the history of what balanced with what in defining prevailing views of costing and pricing; it also reviews how that balance responded to key forces and trends.

The Board-to-Board Religion

In the beginning (at least of this story) was "board-to-board," as practiced through the 1930s.[18] The traces of this beginning remain embedded in the structure of the telecommunications industry of the late 1980s. The political forces that acted on the industry in the 1930s are the political forces that acted on it in the late 1980s— albeit with changed strengths—alongside the powerful attraction of the possibilities opened by new technologies.

Orthodoxy Defined. The justifying philosophy for the board-to-board view of the world was remarkably straightforward (one is tempted to say unconvoluted) by the standards of the 1980s.

An ordinary local call went through the local switchboard, period. An originating interexchange or long-distance call also went through the local switchboard. Only, instead of being routed to another local telephone, it was routed to literally another board, which switched the interexchange call toward some distant destination where this process was reversed.

The distant interexchange board would then route the incoming call to the appropriate local board. Finally, the local board would route the incoming call to the intended telephone.

What could be simpler under these idyllic conditions of yesteryear than to think of costs as being neatly sliced in two at the local edge of the interexchange board?

From the local standpoint, the interexchange board is just another station, and the circuit to the interexchange board just another local loop, with a switchboard instead of a telephone at the end. So the costs of local service naturally include the costs of every piece of personal and local telephone equipment up to the edge of any interexchange board, but not including that interexchange board.

The costs of interexchange boards and of everything that connects them to each other are costs of interexchange service—hence the name "board-to-board" for this approach to defining the costs of interexchange service. Naturally, during the time that this outlook held sway, it was regarded as entirely natural and logical.

It is easy to construct a pricing philosophy in harmony with this costing philosophy. There are three parts to pricing an interexchange call: the price of the two local calls at either end and the price of the interexchange call in between. The whole price to the customer of a long-distance call is then similar to the price of getting from one's home to a hotel in a distant city. One makes separate payments to the cabbies at each end and to the airline. What machinations there might be among the cabbies and the airlines, and how all this gets on the customer's bill, are "details" not to get into in this brief sketch of the way it was.

Costing and pricing went hand in hand, to all appearances, in natural and dissoluble harmony.

Smith v. Illinois Bell: *The Relative Use Heresy.* American courts are like Venus's-flytraps. Their power is nearly absolute over any who put themselves within their reach. But until you're within their reach, the courts—like the flytraps—just sit there. Without a case put before them, they are powerless.

A dispute among stakeholders is what brought costing philosophy before the U.S. Supreme Court in the early 1930s. The most memorable case was *Smith v. Illinois Bell* (1930) (see Figure 18–1). One aspect of the dispute covered the relative scope of federal and state jurisdiction over the telephone industry.

The court took the view that costs must be appropriately divided between state and federal jurisdictions. That finding would have meant little if the envisaged jurisdictional division had followed the three-part cut of the board-to-board costing method. But the court was seen as prescribing another method for the jurisdictional separation of costs: tagging costs according to whether they applied to services under federal jurisdiction or to services under the jurisdiction of one of the states.

The court's language was suitably oracular: "While the difficulty in making an exact apportionment of the property is apparent, and extreme nicety is not required, only reasonable measures being essential, it is quite another matter to ignore altogether the actual uses to which the property is put" (*Smith v. Illinois Bell* 1930).

Eventually, that language came to provide legal underpinnings for what some today call a "joint and common cost" view of local costs. True believers hold it to be obvious to anyone of sound wit and goodwill that local facilities are used for both exchange calls and (interstate) interexchange calls—not for local calls only, as under the classical board-to-board way. For example, Cardullo and Moellenberndt (1987: 40) state that under board-to-board, "all common costs of providing telephone service were charged to local rates" and "toll prices recovered only the incremental direct costs of providing long distance services"; clearly, they subscribe to the prevailing wisdom that viewing costs as common is a matter of natural law rather than a discretionary convention.

Within that convention, the costs of local facilities are joint or common costs of exchange (and state interexchange) and of interstate interexchange services. Or, putting it another way, both interexchange costs and exchange costs run from station to station. In the practice of the 1980s, this would translate as the costs incurred in linking one interface with CPE to another interface with CPE. In the 1930s, of course, the CPE was called a station, belonged to the phone company, and was part of the company's network, not outside it.

From this perspective, the board-to-board view erred in treating the costs at the two local ends as strictly local costs, strictly under state jurisdiction. Instead, since interestate interexchange calls travel over local lines, some local costs properly fall under federal jurisdiction.

Under a different interpretation, the new *Smith* view could have served just as well as board-to-board. From an impartial viewpoint, the argument between board-to-board and station-to-station looks much like the quarrel in *Gulliver's Travels* over whether to open boiled eggs at the big end or the little end. With hindsight and impartiality, it's pure preference. Vague court or legislative language can be translated into whatever outcome you want if the stakeholders agree. Then, as now, fundamental choices about costing are entirely discretionary. Once made, they may have the force of law and hence be very real. A myth turned into law is reality unless and until supplanted by another.

In the 1930s, however, the *Smith* philosophy had no serious muscle backing it. That came only later when those engaging in new practices needed justification and saw it in the new philosophy. Although articulated in 1930, the joint or common cost philosophy— which was eventually tagged "station-to-station" and ultimately supplanted board-to-board—had no immediate practical consequences.

The Lindheimer Case: Heresy Suppressed. Lindheimer v. Illinois Bell (1933) quintessentially stands for a process of juggling costs and prices that has no practical effect on any customer.

The *Lindheimer* case was brought to settle interpretations of the earlier *Smith* case. *Lindheimer* is said to have held that in practice the needs of federalism could be met by putting a federal tag on some appropriate piece of the costs of exchange service, and a similar federal tag on corresponding revenues from exchange service.

The net effect of *Smith* plus *Lindheimer* was word magic— just relabeling. The customer's bill did not change; the company just renamed the reason for collecting the money. A certain proportion of the costs of local plant was attributed to the carriage of interstate interexchange calls and was recognized as being under federal jurisdiction, and then an equal proportion of revenues was assigned to federal jurisdiction. But customers of interexchange services paid exactly what they had paid before, as did customers of exchange services. All that had changed from the earlier ways was the justification for both costs and prices.

Somewhere on the books of each company for each state there was inscribed a number standing for the portion of local costs "separated" from total local costs so as to become the cost of carrying outbound and inbound interstate interexchange calls. Only when it became politically expedient did station-to-station become not only the new cost orthodoxy but also manifest in prices and on bills.

The Station-to-Station Religion: Relative Use Triumphant

> The Department of Justice's consultant's report on competition in the telephone industry, submitted to the divestiture court, stated that, "Allocating truly common costs among the activities they support is a mysterious and fundamentally arbitrary process."
> —U.S. General Accounting Office (1987: 27, cited in Haber [1987: 6.38])

Smith v. Illinois Bell: Apotheosis. Was it good or bad that the judicial cost allocation scheme of *Smith* was for a time made nil in pricing effect by *Lindheimer*?

Fielding questions such as this, although widely practiced, is playing the wrong game. In costing and in pricing there are no absolutes. Consider cost and price definitions as adaptable policy tools instead. They are means, not ends. Neither are they mysteries. At present, as in the past, costing and pricing methods reflect more or less faithfully, albeit with some time lags, the prevailing political balances.

That one among all fairy tales prevails as the law of the land for a time—instead of as a twisted belief to be swallowed against all the evidence of common sense—is an intelligible outcome of a nation's political processes, including, in the United States, the checks and balances among the branches of governments, between states and their federations, and between government and governed.

The question, therefore, is not, is something good, but who are the stakeholders and which of them benefit from a change.[19] The shifting connection between costs and prices is significant only in terms of before-and-after comparisons of *relative* losses or gains from a change, as perceived by stakeholders. The netting out of these gains and losses is a political question in the best sense of the word.

It is historical fact (see Figure 18–1) that in the 1940s—more than a decade after its de jure canonization as a cost standard in the law, and after a decade of de facto irrelevance to daily living (especially to any perceptible pricing)—the vague idea of relative use enunciated by the Supreme Court in *Smith* began its trans-mogrification into the station-to-station philosophical piling that, although rotting and shaking by the late 1980s, still held up over the swirling waters of a changing world the piers of actual costing and pricing practices.

In the early 1940s, the political, economic, technological, marketplace, and other constellations of the day came into a conjunction that found handy justification in substituting the joint and common cost ideas of *Smith* for the "local is local and interexchange is interexchange and the twain meet only at the interexchange board" ideas of board-to-board.

By the early 1950s, the essentials had come into place with the promulgation of the first *Separations Manual*, following the sign-on of the independents to the settlements process. Not until 1970, with the adoption of the modifications (known as the Ozark Plan) to the *Separations Manual*, did the joint and common costing principles achieve their apotheosis when state and federal separations and settlements were harmonized by the acceptance of a common costing and pricing scheme.[20]

Although embodied in several formulas, the costing principles were epitomized within the industry by two notorious factors in one of the formulas. The first of these factors was the subscriber plant factor (SPF). The SPF was calculated from other factors—including the subscriber line usage (SLU), whose definition was many-layered.

The associated pricing system hardly nodded to the costing system, except that, in the aggregate, total costs equal total revenues within each state and within the federal jurisdiction. Once a suitably sized chunk of costs had been moved to the federal jurisdiction by the separations process, pricing in the states generally went as sketched in Figure 18–7.

The industry and the regulators achieved consensus on a process of residual ratemaking. That meant that when an intrastate price increase loomed up, the first line of defense was to shift costs to the federal jurisdiction. Within a state's jurisdiction, state interexchange or vertical services were the pricing target of first choice. Second choice was directory advertising. Only as a last resort was an exchange price hike considered.

Ironically but not uncommonly, when the Ozark Plan started in 1970 the underpinnings of the order that supported the costing practices blessed by the *Separations Manual*, and the pricing built on them, had already been thinned out.

In 1957, the *Hush-a-Phone* decision (FCC 1957) had pulled some supports out from under the telephone industry's ban on what

FIGURE 18–7

» *Precompetitive Pricing Method Within States.*

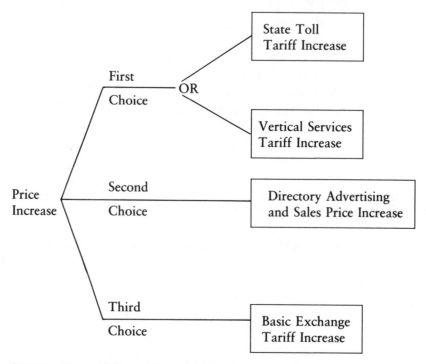

© 1988 President and Fellows of Harvard College. Program on Information Resources Policy.

it called "foreign attachments" to its network. In 1960, *Above 890* (FCC 1959; FCC 1960) allowed any large customer to build his own facilities along patterns that the Communications Act of 1934 had reserved for privileged "public service" entities such as railroads, airlines, and brokerage houses. And in 1970, the Federal Communications Commission's (FCC's) first MCI decision (FCC 1970) allowed MCI to begin competing with AT&T with what was said to be a limited service designed to meet the interoffice and interplant communications needs of small businesses.

The business of the largest interexchange customers and the use of the highest volume routes of the traditional industry could no longer be taken for granted by either the industry or its regulators. The entry of competitors not tied to the separations and settlements process threatened the stability of the process.

From SLU to SPF Under Politics and Policy-Supporting Monopoly. *Smith* enshrined station-to-station theory in 1930. *Lindheimer* nullified any practical consequences of the new theory in 1933. The shift from board-to-board *practices* to station-to-station *practices* began only in the 1940s; the practice peaked by 1970, and it continued with much more vitality than the Cheshire Cat's grin well into the late 1980s, decades after its own *theoretical* (or judicial or economic) pilings had begun to rot.

Figure 18–8 shows secondary shifts that occurred within the prevailing station-to-station costing method. An increasing shift of costs from state to federal jurisdiction had become automatic and accelerated as growth in SLU accelerated, amplified by growing multipliers.

The milestones in Figure 18–8 mark the progression from SLU to SPF, as one important measure of the "relative uses" of *Smith.*

FIGURE 18–8

» **The Road from SLU to SPF.**

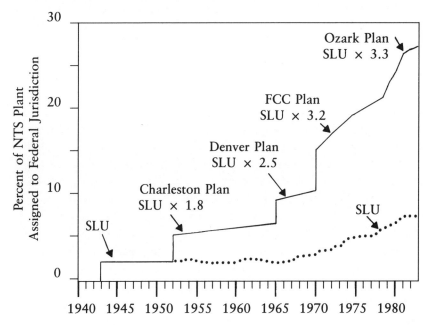

SLU is itself a discretionary choice (Cunningham 1917: 238). SPF was once justified as a necessary reflection in costs of deterrent effects of usage-sensitive interexchange pricing.[21]

There is little disagreement as to the effects of the practices supported by the station-to-station philosophy and expressed in a variety of administrative mechanisms. Costs that hitherto had been reckoned among the costs of exchange service were henceforth reckoned among the costs of interstate interexchange service. In and of itself, this would have been neither new nor significant; witness the immediate aftermath of *Lindheimer.*

What was significant was that on this occasion costing practice made a real difference for one simple, good, and sufficient reason: politically, there was enough of a supporting consensus favoring the policies that the costing practice was meant to support, and not enough opposition to those policies. Administrative and technical practice was, as it should be, the servant of its political policy-setting masters.

This was, in brief, the outcome of the politics of rural and populist political dominance—with maneuvering room left for the more subtle shadings of urban and business politics.

In the aggregate, the shifted costs began to appear on the bills of the customers who bought interstate interexchange services. An equivalent aggregate amount failed to appear on the bills of those customers who bought intrastate services. Within states, there was a general tendency to flow more of this pricing benefit to prices for exchange services rather than to the prices for intrastate interexchange, according to the procedures illustrated in Figure 18–7. In their heyday, these practices were, if not enthusiastically supported by an explicit political consensus, at least without muscled detractors.

The interexchange users who paid for the shifted costs originally were a small subset of exchange users—business, and especially, large business. Granted, their customers or owners ultimately paid these costs, but these costs were (and in many instances remain to this day) such a small portion of their total costs as to be unimportant. By the 1960s, in contrast, interexchange usage had begun to acquire an increasingly widespread, growing, and more enfranchised household constituency.

In this period, the unit cost for long-haul facilities declined with a robustness apparent in any reasonable cost measure. Rather than decrease long-haul (read, interstate, federally regulated) prices, revenues recovered from the interstate services covered costs shifted from intrastate services, thereby preventing price increases there.

Ultimately, the communications managers of the Fortune 500 began to present testimony on behalf of organizations whose top managements remained as oblivious to the largely invisible telecommunications costs as before. The political action of these telecommunications managers helped to destabilize the "station-to-station under monopoly" order of things in an environment where no one else, except the monopoly suppliers, either cared much or could do much. Rural populations had by then migrated into the cities, and the Supreme Court's "one person, one vote" decision in *Baker v. Carr* (1962) had grown teeth.

The foregoing facts are generally accepted. There were, and are, vehement disagreements over the interpretation of these facts.

Some hold that the shift to station-to-station costing was a long overdue recognition of the true costs of services. Others hold that it was, pejoratively speaking, an unfair cross-subsidy of exchange services by interstate interexchange services. Still others argued that, to the contrary, exchange services were cross-subsidizing interexchange services.

None of the above arguments makes any sense. All presuppose that there is some ideal, correct, true cost. Such arguments fail to recognize the stubbornly discretionary character of practical costing and pricing. The formula is everything.

For many of the stakeholders—administrative agencies and telephone company managements among them—one of the beauties of the pricing practices of the station-to-station heyday was that, except for a rough equality of costs to revenues at the "costs equal revenue requirements" level of the prevailing rate base/rate-of-return method of regulation, the costing method was not seen to reflect any particular pricing method.

The resulting latitude permitted flexibility in adjusting prices to local political variations. Where there was demand for relating prices to costs, suitable costs were invented to justify prices. Economic theory in the service of one side or another of rate cases merely put a veneer of economic "truth" over the wood of political compromise.

One took his professional life in his hands in those days for suggesting that the exercise was valuable for building political smoke screens, but intellectually empty. Smoke screens are seen as more respectable, and maybe more effective, when they are also grounded in "sound principles." Hence, the bull market for compliant philosophies in changing times.

When pleas for favorite orthodoxies are abandoned in favor of examining the match between the instruments for administering policy and the policy of the day as defined by the political system, many layers of clouds dissipate and lunatic-seeming rituals fall into place.

SLU and SPF emerge as the earthly incarnations of the Supreme Court's vague "relative use" doctrine in *Smith*. Relative station-to-station–use, joint and common cost allocation rules make perfectly good sense in this light. In the context of a consensus for monopoly— or at least, only unsuccessful assaults on it—and a consensus for increasing telephone penetration into U.S. households from 45 percent in 1943 to 95 percent and more in the 1980s, the administrative tool served the policy masters well.

Once the masters had changed their minds, small wonder that the tools of the old order became suspect as the 1970s waned into the 1980s. It may therefore seem all the more remarkable how little has changed in practice over the last decade, rather than how much.

But that is the norm. In achieving compromise, American political, administrative, and judicial processes typically introduce change gradually, with an initially almost imperceptible shading of the old into the new.

Table 18–4 illustrates this point and more. It displays the rationale for the SPF formula, frozen and discredited by the late 1980s, in the warm glow of its birth at the end of the 1960s. Each element is accounted for by the policy objective that it served.

Specifically, the factor 0.85 (fifth line of Table 18–4), so mysterious and so technical in appearance, made the transition from one costing method to another an unobtrusive and painless event. The 0.85 factor produced so close a match with preceding practice that nothing at all happened at the moment of transition. The changes came later. Such are the ingredients of a masterpiece of the administrative arts. Figure 18–8 shows how smooth the transition to SPF was compared to earlier, clumsier transitions.

The masterpiece was seen as junk, of course, once the policy it served had died. In the turmoil of the late 1980s, with a greater number of active stakeholders than before, especially among suppliers, and with no consensus in sight, small wonder that the search for satisfactory tools continued. But the junking process was just as smooth as the earlier polishing process. There was no sudden death for SPF: at the moment of truth, the live allocator with wide LEC-by-LEC swings around an average value of 26 percent was tranquilized into a gradual transition toward a fixed 25 percent allocator for every LEC.

» **Table 18–4. Policy Ends Met by Means of the SPF Formula.**

Means	Ends
SPF = (0.85 + 2 × CSR Ratio) × SLU	Sets up a well-run and honest program. Single formula used for federal cost allocation, for federal settlements, and for state settlements.
(0.85 + 2 × CSR Ratio)	The SLU multiplier, by replacing the SLU additive of earlier formulas, was proclaimed as marking the end of the cycle of continuous revisions of federal cost allocation method. As use (SLU) increases, the cost allocation (SPF) increases.
2	Replaces the additive of the previous cost allocation plans which essentially doubled SLU.
CSR Ratio	Adjusts cost allocation (state-by-state, company-by-company) based on the relationship of prices and lengths of haul. Favors some states over others relative to earlier federal settlement plans. Enables state settlements to use same formula as federal settlements.
0.85	Provides a smooth transition from the previous separations plan. Also provides a straight SLU additive.
SLU	Satisfies the court requirement to tie cost allocation to "actual uses." Adjusts cost allocation by each company's interexchange traffic.

Cost-based Pricing: Occult for the 1980s

"All right," said the Cat; and this time it vanished quite slowly, beginning with the end of the tail, and ending with the grin, which remained some time after the rest of it had gone.

"Well! I've often seen a cat without a grin," thought Alice, but a grin without a cat! It's the most curious thing I ever saw in all my life!"

—Lewis Carroll, *Alice in Wonderland*

From SPF Back Toward SLU Under Politics and Policy Supporting Competition. As Figure 18–1 shows, the principal elements of the old costing order are in transition—toward where we know not.

Not only are more supplier stakeholders in the field than before, but government intervention in the late 1980s is also more varied, more extensive, and more Byzantine than when the vogue for competition and deregulation began in the 1960s.

The traditional players in the telecommunications game were the Bell System and the independents. Then one Bell System abruptly became eight players on 1 January 1984. Numerous other carriers entered the field, each vocal, even if not potent. The computer and the consumer electronics industries patrolled the no-man's-land at the shifting borders. The newspaper industry, scared by the potential it saw for incursions by electronic yellow pages into its classified advertising field, unfurled the First Amendment banner over the astonished "common carriage" field. And reminiscent of the shippers who became railroaders in the late 1800s, end-users with excess capacity on their private networks to sell also came into the market.

What many in the formerly traditional industry hoped might be a "flash-cut" to a new deregulated order unfolded instead as a typically slow-motion political adjustment designed to smooth out shocks on the way to some still unseen, hence unformulated and unformulable, consensus.

It is therefore a fact of life that the new, somewhat more competitive, and much more regulated world of the late 1980s retained key elements of a deceased political consensus.[22]

Of course, since the Supreme Court's 1962 decision favoring "one person, one vote," political power has moved toward the urban concentrations, which produce high-volume traffic, and away from sparsely settled rural areas, where the "higher costs" are.

What "higher costs" means leads to another set of wheels within wheels. "Higher cost" and "rural" areas include places like New York

City, Chicago, and Los Angeles suburbs and exurbs, not just Appalachia and Western ranches or deserts. Real boonies may entail higher expenditures or maybe even higher costs per subscriber line; but they may also entail older plant that is fully depreciated, that is, whose costs may be almost fully recovered. Or new plant may have been built with low-cost capital such as Rural Electrication Administration (REA) loans. Under such circumstances, local service prices might be lower in those "high-cost" areas than in "lower cost" areas.

Meanwhile, as Figure 18–1 shows, the SPF was frozen on 1 January 1983. During the debate over the freeze, passionate stakeholders could be seen sporting buttons with the slogan, "SPF to SLU in '92." The SPF was slated to phase down to 25 percent over eight years, beginning on 1 January 1986, except that no company's SPF could go down more than 5 percent in a year. Since the maximum SPF is 85 percent, some companies—or more accurately, some study areas or states—may take twelve years to reach 25 percent.

None of the twelve-year companies are former Bell System companies. The spreading out of relative pains and joys over a decade or so is a classic technique of political compromise (Borchardt 1976).

The closeness of the 25 percent figure to the national average of 26 percent also means that the effect of this change on AT&T was, for better or worse, close to nil.

The very costs apportioned by SPF were not, in the late 1980s, any more stable than the frozen and evaporating SPF itself. Costs were invented, not discovered, according to the rules in place during the heyday of the SPF's Ozark Plan.

Costs were recorded according to the rules governing the uniform system of accounts (USOA). The rules for jurisdictional separation laid down in the *Separations Manual* and in Part 67 of the FCC rules were then applied to those entities to produce the costs subject to federal jurisdiction. In addition, the investment costs had a rate-of-return factor applied to them. Similar processes are applied to USOA accounts to construct the other types of costs.

Under the *Separations Manual*, each band was then split into state and federal costs according to some formula. The Ozark Plan SPF formula of Table 18–4, when applied to various components of plant, yielded the federal share of the costs. These costs, built by very human hands, are a monument to discretion in determining federal cots.

That discretion continues to be exercised, altering the very foundations of the costing process. In the late 1980s, the USOA was being re-worked while the house it supported continued to be lived in.[23] Also re-worded were the rules in the *Separations Manual* and in Part 67 of the FCC rules[24] that defined how costs as recorded in the accounts of the USOA were to be picked out and eventually recombined.

Since 1980, other discretionary moves have combined with these to alter the categories of plant. Some of these have been structural.[25] For instance, the non–traffic-sensitive (NTS) local dial band was eliminated on 1 January 1988. This move is one among several changes considered in a proceeding on central office equipment (COE) separations rules. There is no concomitant physical change in any network—just in the way costs, and indeed functions, are defined for separations and interstate access charge purposes. Meanwhile, the SPF has been frozen, and other formulas are under continuous alteration.

In all this, the basic political process remains: in the 1980s, as before and most likely afterwards, the formula is everything.

> The grand old Duke of York,
> He had ten thousand men.
> He marched them up to the top of the hill,
> Then he marched them down again.
>
> —Folk song

Lindheimer Lives!: Heresy Triumphant in the SLC. The political foundation that aligned station-to-station costing practices with pricing practices began to erode when *Above 890* enfranchised large and concentrated end-users to build their own networks if they chose (and later even to resell excess capacity to others), or to give their business to others enfranchised by *MCI* and its successors. Simultaneously, small but concentrated urban end-users were increasingly enfranchised by the slow but steady implementation of *Baker v. Carr* and the concomitant reduction in the power of rural interests.

With increasing interexchange competition, and with divestiture, the residual AT&T lost some of its incentive to play the settlements side of the station-to-station–justified pricing game: in the old days, the settlements process had passed to the BOCs the separated cost–justified revenues that AT&T collected, on the basis of cost separations, from its interstate interexchange customers (see Figures 18–2 and 18–3). These revenues settled in what amounted to the left-hand pocket instead of the right-hand pocket of one family. After divestiture, the pockets belonged to strangers who might become competitors. And payments to the BOCs and the independents put an upward pressure on interexchange pricing, which, unless offset by other factors, disadvantages an IC making those payments relative to ICs not required to make them, or required only to pay less.

Settlement payments by the Bell System to independents had always been a highly controversial area. The Bell System acquiesced in them as part of the political compromises that drove the transition from board-to-board to station-to-station methods.

The continual rise in interstate SLU, already noticeable from the 1950s through the 1970s, further accelerated in the 1980s. Multiplied by SPF, this increase was leading toward the situation in which AT&T paid about 50 percent of its revenues over to other companies, the LECs, who had come to be described as its suppliers of access services.

More important in the overall political balance, perhaps, was the fact that by the 1980s this revenue came from a broad base of end-users. These had come to include not only the traditional business users but new urban dwellers apt to use interexchange service to keep in touch with the folks back home. The political perception had shifted toward a sense that interexchange users, not exchange users, were paying more than a "fair share" of the costs.

Besides, the goal of universal service within the United States had been reached in terms of the practical political standards of the time, at least for voice services, which old telephone hands used to call POTS (plain old telephone service). Although revered by some, the idea of universal digital, data, or voice/data services was ahead of its time in practical market or political terms.

The outlook had become favorable, in short, for a reversion to the board-to-board religion, which would sanctify interexchange customers paying only for what had, by then, become comparatively negligible investments in long-haul plant. But this was not to be—at least not by the late 1980s, given the residual political strength of the beneficiaries of station-to-stationism and the professional respectability meanwhile acquired by concepts of joint and common costs.

Return to full-blown board-to-board costing and pricing was not in the political cards of the 1970s and the early 1980s. The result was a kind of reverse *Lindheimer*. In this swing of the pendulums, the costing philosophy and practice both stayed put, but the failure to change costing practice was partly compensated for by a real switch in pricing methods.

As the FCC's Access Charge Plan gradually went into effect in the mid-1980s in the face of vocal congressional objections but no legislative action, the enemies of station-to-station thinking were luckier than the enemies of board-to-board had been in the early 1930s. *Lindheimer* had essentially nullified the practical pricing effects of an anti–board-to-board change in costing theory and practice by keeping the older ways alive in pricing practice, although dead in costing theory. The Access Charge Plan also nullified the practical pricing effects of *no* anti–station-to-station change in costing philosophy by killing some of the older pricing practices, although sparing their lives in costing theory.[26]

The Access Charge Plan aimed at grafting a board-to-board–era pricing method onto the preexisting station-to-station costing practice. Hence the continuing presence of a station-to-station costing grin as part of the station-to-station pricing cat disappeared. That this succeeded even in part is a tribute to political, administrative, and judicial artistry as consummate as *Lindheimer* and the Ozark Plan were in their heydays.

In the mid-1980s, it had been more fashionable than in the past to seek a more direct tie between prices and costs, more because of the rhetoric of some increasing competition than because of the realities of competition. But even in the sectors of the American economy that are the most competitive and the least subject to government intervention, workaday prices have little relation to hypothetical and immeasurable economic costs. Fairy tales abound for internal incentive, Internal Revenue, and other diverse purposes; in those realms, too, the formula is everything.

Stuck with both the station-to-station costing grin and the competition-induced fashion of tying pricing more directly to costs than under residual pricing, the access charge planners simply carved suitable costs from the grin (Lemler 1987: 38) and then tied them to pricing.

The starting point of access charge pricing was separated costs, as defined by existing station-to-station cost separations (including the USOA and a frozen but evaporating SPF).

The process then assigns costs based on old-fashioned methods to categories with new fangled access names, negotiated in proceedings with newfangled stakeholders in them. Access pricing is based on these relabeled and regrouped costs. As before, economists protest about politicization, but they work with these costs just the same.

The pricing method favored by the access charge planners in the FCC would have resulted in a pricing process remarkably like the *Lindheimer*-era pricing method. End-users would have paid for the interstate part of the wires connecting them to the nearest switch, whatever actual use they might make of them, on the ground that these wires all served exchange purposes, including occasional access to the interexchange network.

This view accomplishes in *pricing* terms what board-to-board accomplished in *costing* terms, and what *Lindheimer* maintained by nullifying the pricing consequences of a change in costing. In the FCC's Access Charge Plan, certain costs remained colored federal, thereby satisfying the *Smith* strictures as to cost allocation. But a federal mandate that these costs be paid by all end-users ensures, in a reverse-*Lindheimer* twist, that the federal costs will once again be paid by all end-users and no longer by the users of interstate services only. A pricing change here nullifies a nonchange in costing.

Of course, Congress understood as well as the FCC, and perhaps even better, that the formula is everything. And so, under pressure from Congress, the Access Charge Plan as originally formulated was modified and, as of the late 1980s, became stalled between station-to-station–style and board-to-board–style practical effects on which end-users pay for what.

The projected transition for recovery of non–traffic-sensitive costs from station-to-station–style pricing to board-to-board–style pricing will occur over the next several years.[27]

In 1986, the process had stalled, with only $2.00 of an originally proposed $4.00 in effect as the (board-to-board–style) SLC for residential customers and single-line business customers. Six dollars was the ceiling for multiline business customers, with actual prices linked to a costing formula. The (station-to-station–style) CCLC also remained in effect—combined with the effects of other costs not detailed here.

By late 1987, the residential SLC had moved to $2.60 and was slated to go to $3.50 by 1 April 1989. And in the continuing evolution of the "new" board-to-board pricing, the idea that the ICs might "always" have to continue to pay for access to subscriber lines had, for the moment, gained explicit recognition by the FCC.

All of the other discretionary elements discussed in this paper, along with many more not touched on, remain controversial—mainly within the state and federal administrative agencies but also in the courts and in Congress—as products and services, along with their costs and their prices, continue to evolve within the political process and the marketplace.

» References

Aulik, Jaak. 1987. *Financial Structures in Competitive Telecommunications: An International Overview.* Publication P–87–2, Harvard University, Program on Information Resources Policy.

Baker v. Carr. 1962. 369 U.S. 186.

Borchardt, Kurt. 1976. *Toward a Theory of Legislative Compromise.* Publication P–76–4, Harvard University, Program on Information Resources Policy.

Cardullo, J. Patrick, and Richard A. Moellenberndt. 1987. "The Cost Allocation Problem in a Telecommunications Company." *Management Accounting* (September): 40 ff.

Communications Act of 1934, Title I, Section 3(r), 48 *Stat.* 1066; codified in 47 *USC* 153(r).

Cunningham, William J. 1917. "The Separation of Railroad Operating Expenses Between Freight and Passenger Services." *Quarterly Journal of Economics* 31 (1917): 238 ff.

Epstein, Samuel M. 1985. *Behind the Telephone Debates—4: A Conceptual Framework for Pre- and Post-Divestiture Telecommunications Industry Revenue Requirements.* Publication P–85–7, Harvard University, Program on Information Resources Policy.

Federal Communications Commission. 1957. *Hush-A-Phone Corp. v. AT&T Co.,* et al. FCC Docket No. 9189, *Decision and Order* (21 December 1955): *Decision and Order on Remand,* 22 FCC 112.

———. 1959. *Allocation of Frequencies in the Band Above 890 Mc.* FCC Docket No. 11866, *Report and Order,* 27 FCC 359.

———. 1960. *Memorandum Opinion and Order.* 29 FCC 285.

———. 1970. *Microwave Communications, Inc. (MCI).* FCC Docket No. 16509, *Decision,* 18 FCC 2d 953 (1969); *Memorandum Opinion and Order,* 21 FCC 2d 190 (1970).

———. 1978. *MTS and WATS Market Structure Inquiry.* FCC CC Docket No. 78–72.

Gabel, Richard. 1967. *Development of Separations Principles in the Telephone Industry.* East Lansing: Michigan State University Press.

Gartner Group. 1987. *Point-to-Point.* Stamford, Conn.: Gartner Group (2 October).

Huber, Peter W. 1987. *The Geodesic Network: 1987 Report on Competition in the Telephone Industry.* (Huber Report) Washington, D.C.: U.S. Department of Justice, Antitrust Division.

Lemler, Mark L. 1987. *The FCC Access Charge Plan: The Debates Continue.* Publication P–87–8, Harvard University, Program on Information Resources Policy.

Lindheimer v. Illinois Bell Telephone Co. 1933. 292 U.S. 151.

Masoner, Jeffrey A. "Alternatives to Rate of Return: Stakeholders and Positions." Mimeo (undated), Harvard University, Program on Information Resources Policy.

Modification of Final Judgment (MFJ). 1982. *U.S. v. AT&T* 552 F. Supp. 131 (D.D.C. 1982), aff'd. mem., 103 S. Ct. 1240 (1983).

Sichter, James W. 1977. *Separations Procedures in the Telephone Industry: The Historical Origins of a Public Policy.* Publication P–77–2, Harvard University, Program on Information Resources Policy.

———. 1987. *Profits, Politics and Capital Formation: The Economics of the Traditional Telephone Industry.* Publication P–87–7, Harvard University, Program on Information Resources Policy.

Silberberg, Jay L. "Alternative Telecommunications Costing Methods." Mimeo (undated), Harvard University, Program on Information Resources Policy.

Smith v. Illinois Bell Telephone Company. 1930. 282 U.S. 150–151.

U.S. Congress. House of Representatives. Committee on Government Operations. Subcommittee on Government Information. 1983. *Hearings: Access Charge Impact: The Impact of the FCC's Telephone Access Charge Decision.* 98th Cong., 1st sess. (18 May, 22 June, 21 September, and 27 September).

U.S. General Accounting Office. 1987. *Telephone Communications: Controlling Cross-Subsidy Between Regulated and Competitive Services.* RCED–88–34. Washington, D.C.: GAO (October).

U.S. v. Western Electric. 1987. Decision of the U.S. District Court for the District of Columbia (10 September). Mimeo.

Weinhaus, Carol L., and Anthony G. Oettinger. 1988. *Behind the Telephone Debates.* Norwood, N.J.: Ablex Publishing Corporation.

» Notes

1. This paper draws on the contributions of many participants in the Harvard Program on Information Resources Policy. For the most extensive and most recent help, I am especially grateful to Jeffrey A. Masoner, Jay L. Silberberg, and Carol L. Weinhaus.

   Thanks are also due to Mark Lemler, Jeffrey A. Masoner, John McGarrity, and Jay L. Silberberg for reviewing previous drafts. These reviewers and the program's affiliates are not, however, responsible for or necessarily in agreement with the views expressed herein, nor should they be blamed for any errors of interpretation.

2. See, for instance, the decision in *U.S. v. Western Electric* (1987). In Part VIII ("Transmission of Information Services"), Section F ("Necessary Infrastructure Components"), the court proposes a compromise between newspaper and telecommunications interests. But this compromise meanders into the no-man's-land between telecommunications and computer industry interests fought over mainly in the FCC's Computer Inquiries I, II, and III. This compromise also makes excursions across the exchange/interexchange boundary (see especially note 308 of the decision), which, elsewhere in the decision, is presented as the centerpiece of controversy within the telecommunications industry.
3. For a fuller background, see Weinhaus and Oettinger (1988). Under way at the Harvard Program on Information Resources Policy are detailed studies of alternatives to current costing methods and current approaches to price regulation (Silberberg [undated] and Masoner [undated]).
4. A detailed discussion of how exchange and interexchange services are defined is given in Weinhaus and Oettinger (1988: ch. 12). Historically, interexchange services were called "toll" services, and exchange services were called "local" services, but "interexchange" and "exchange" are used exclusively throughout this paper.
5. By the end of 1987, "subscriber line charge" had become the prevalent term for the flat-rate, end-user access charge. Earlier terms for the same concept include "customer access line charge" (CALC) and "end-user charge" (EUC).
6. The specifics of access charges have been laid down in a series of FCC decisions, mostly within FCC (1978). Details are given in Weinhaus and Oettinger (1988) and in Lemler (1987). Congress has also inquired into the matter. See, for example, *Access Charge Impact* (1983).
7. As of early 1989, the Program on Information Resources Policy's work in progress was able to break down the $4 billion into roughly $3 billion for SLCs and $1 billion for other access charges. The latter include charges paid by ICs such as Alltel, charges for inter-LEC interLATA corridor traffic, payments under BOC-independent arrangements that succeed state settlement agreements, and the charges for special access services to businesses. Special access includes services that carry high-volume traffic directly from the businesses to an IC's point of presence (POP)—one form of so-called service bypass—and services by LECs directly to businesses, the so-called customer-ordered LEC access (COLA).

8. Following divestiture, six of the BOCs created separate subsidiaries for directory services. Bell Atlantic alone retained directory services within its regulated telephone companies. In the late 1980s, this matter was under review in the courts of several states, Arizona among them.

9. In the Program on Information Resources Policy's work in progress, the treatment of certain revenues of both ICs and LECs remains fuzzier than it ultimately needs to be. Specifically, further disaggregation and categorization of certain nontelephone revenues will alter both the totals and the proportions in Figures 18-2, 18-3, and 18-4. The overall static impression will not change materially, since the amounts in question add up to less than 10 percent of the total. The flux in that 10 percent is important, however, to understanding the dynamics of costing and pricing strategies.

10. For details, see Weinhaus and Oettinger (1988: 60, Fig. 8.6). For a 1980 baseline, see also, Sichter (1987: 107, Fig. 3.7[a] especially).

11. Access charges are also levied for *intra*LATA access, although there are a variety of exceptional cases.

12. Notes 6 and 8 above explain the sources of this fuzziness.

13. Data as of early October 1987 suggest that the term "substantial portion" means about 75 percent (Gartner Group [1987: 6]).

14. See note 7.

15. Estimate from Gartner Group (1987).

16. For details on the international aspects, see Aulik (1987).

17. Some confusion in the terminology has been injected by the courts. The text of the 1982 consent decree of the court in *U.S. v. Western Electric* (also referred to as the Modification of Final Judgment) uses the term "interexchange" in a way that the parties and the court defined to be synonymous with "interLATA." The LATA is a concept that was itself tailored to the exigencies of the MFJ. The older usage of "interexchange" by the industry is broader. This usage holds interexchange services in contrast to exchange services, as defined in the Communications Act of 1934 (see note 6 above). It makes sense, therefore, to refer to *intra*LATA *inter*exchange services. In late 1987, intraLATA interexchange services were authorized by eighteen states, including Florida and South Carolina, where authorization had a more limited scope than in the others.

18. For details on the history up to the 1930s, see Sichter (1977); see also, Gabel (1967).

19. Sichter (1977) has examined these issues.

20. In the late 1980s, actions on the *Separations Manual* and on Part 67 of the FCC's rules, which legitimate the manual, have been

taken mainly under the FCC's CC Docket No. 80–286. See Weinhaus and Oettinger (1988) for the earlier history.

21. The FCC said that its purpose was "to compensate for the deterrent effect on actual use in each state of the nationwide interstate toll rate schedule." The idea was that usage would have been higher if people had not had to pay prices tied to distance. The FCC did admit that "the deterrent effects on toll use of subscriber plant resulting from the structure of toll rate schedules cannot be quantified with exactitude. We are, thus, required to use our best judgment . . . as to the weight that should be accorded to these effects." These specimens are from FCC Docket No. 16258: the first at 9 FCC 2d 960 (1967), the second at 9 FCC 2d 30 (1967).

22. It is also a fact of life that the deceased political consensus remained alive in the minds of a still potent rearguard, exemplified by men like Rep. Jamie L. Whitten (D-Miss.), Chairman of the House Committee on Appropriations, Chairman of the House Subcommittee on Rural Development, Agriculture, and Related Agencies, and longtime champion of the REA, the guardian angel of the smallest independents.

23. The USOA is prescribed in Part 31 of the FCC's rules (47 CFR Part 31 [1982]). Proceedings to amend the system were under way in the late 1980s in FCC CC Docket No. 78–196. The revised system (USOAR) is in Part 32, hand in glove with conformed separations in Part 36.

24. See note 20.

25. The transition is examined in Weinhaus and Oettinger (1988) and, in much greater detail, in Epstein (1985).

26. Details are in Weinhaus and Oettinger (1988) and, in greater depth, in Lemler (1987).

27. Details are in Lemler (1987).

19

» *Options for the Future: Cost Relationships*

Ron Choura

» *Current Joint Boards*

Federal/state joint boards are an administrative procedure set up under the Communications Act of 1934 for determining how costs will be allocated to the proper federal or state regulatory jurisdictions. This paper discusses the federal/state joint boards in terms of what is going to affect some of the decisions being made on pricing and of what problems may arise in the splitting of costs and in the jurisdictional balance between state and federal commissioners.

Currently, federal/state joint boards are created by the Federal Communications Commission (FCC). There are three joint boards in existence, and three related FCC dockets: 80–286, 85–124, and 83–1376. The first docket is concerned with all jurisdictional separations changes on a wholesale basis. Currently, it focuses on specific issues. The second docket is concerned with feature groups—bundled services that are purchased right out of a tariff. This docket is examining Feature Groups A and B; specifically, how to allocate costs between federal and state jurisdictions for Feature Groups A and

B circuits subscribed to by customers. This is clearly a major pricing problem, but diminishing in importance as time goes on. The last docket, 83–1376, while focusing primarily on Alaska and Alaska's problems in cost and cost allocations, has significant, far-reaching implications for pricing in the long run. It is basically concerned with a rural subsidy issue—whether small telephone companies and rural America will continue to be able to be part of the nationwide rate integration for toll rates, or whether toll rates will begin to be deaveraged. The docket will determine the policy on that issue.

» Past Decisions

The past joint board separations changes affect pricing from the telephone instrument all the way through to the interexchange plant. The joint board will continue to make changes in the marketplace, as well as in cost allocations. The joint boards have dealt with customer-premise equipment (CPE) and inside wiring: the non–traffic-sensitive outside plant. They have recommended changes in the separations formula to shift from the old subscriber plant factor (SPF)—which had upwards of 86 percent of the plant allocated to the interstate jurisdiction—to a 25 percent gross allocator to be phased in over a period of eight years. We're exactly halfway into that phase-in. Clearly that phase-in effect has major implications for pricing, since prices must reflect jurisdictionally allocated costs. Because this policy only dealt with the interstate jurisdiction, many of the state commissions must develop similar policies in their own jurisdictions. State commissions may adopt a 25 percent gross allocator, subscriber line usage (SLU), a weighted SLU.

What will be the final outcome is unclear, but all of these policies have major implications for the costs assigned to the local exchange carrier's (LEC's) local revenue requirement, and for how the LEC will price services in an open network architecture (ONA) environment. In implementing ONA, the problem is how to recover the cost of the local loop when that loop is unbundled. How the costs are allocated will make a significant difference. These are only some of the major concerns in jurisdictional allocations that are forthcoming.

Additionally, the development of high cost funds (HCFs) has added a different dimension to pricing. In a competitive marketplace, HCFs become a very critical issue, whether or not HCFs for local companies can be maintained. The joint board has looked at HCFs four different times—and has come up with four different recommendations on allocations formulas for HCF distribution. While the joint board has made what is termed a final recommendation, with all the pricing changes going on with the new system of accounts, the new separations manual, and the implementation of many changes in the last year, that HCF formula is probably going to be revised, since the distribution that was settled on in this particular mechanism has totally changed. Where the money is coming from will be different. One additional point is that how the revenue for HCFs are recovered will change in January 1989. This will affect prices.

Direct Assignment of WATS

The joint board recommended on two occasions direct assignment of WATS, which the FCC finally adopted; but a year ago the FCC implemented bi-jurisdictional WATS. Basically, the costs of one WATS access line that carries both state and interstate is common to both jurisdictions. This clearly throws into question the jurisdictional direct assignment of WATS loops and forces the joint board to revisit this issue. The issue is currently before the board. How it ends up allocating WATS costs will change both allocated costs and pricing decisions.

Changes in the allocations factor for central office equipment (COE) have occurred. A major portion of a company's cost and its allocation are COE costs. The joint board made recommendations in March 1987 to allocate COE based on dial equipment minutes of use (DEM). However, whether the DEM formula is the most applicable surrogate means of allocating central office costs is debatable. A second allocator—switched minutes of use (SMOU)—is also being considered.

Fundamental, philosophical differences exist betwen the two methodologies. There is currently no evidence as to which is correct. The joint board is now considering the issue, and a recommendation is expected sometime next year. Any changes would make a significant difference in the amount of central office costs allocated to the interexchange business, especially if the interstate method of allocation is mirrored in the instrastate jurisdiction. Currently, the difference between DEM and SMOU is about a factor of 1.3—a significant difference in the allocation of costs.

Another change to jurisdictional cost allocations being examined concerns the circuit equipment that connects the COE to the outside plant trunks. In March 1988, the joint board recommended that LECs and interexchange carriers (ICs) change the way their costs are allocated. The factor has been frozen for ICs. However, that freeze may be changed by the joint board in Docket 83–1376 (the review of Alaska's problems with cost and cost allocations).

The last major change is one of conformance to the new system of accounts recommended by the FCC. The FCC changed from the old uniform system of accounts Part 31, to the new Part 32, effective January 1988. The joint board has recommended a new separations manual to be used with the new Part 32. While the joint board has done modeling techniques to ensure that there will be minimal jurisdictional revenue shifts, it cannot be certain that revenue shifts will not occur. In fact, even the telephone industry does not know how the mapping from Part 31 to Part 32 will occur. The joint board still has major questions and clarifications to address over the next couple of years. The mapping and allocation will change the bottom line—that is, the jurisdictional cost allocations—and hence, will change the prices of the various services.

The joint board recommended some major changes in March of 1987. These changes were to subscriber line charges, life-line funds, HCFs, pooling, revenue accounting, Account 662, central office circuit equipment, local dial switching equipment, and a new conformance separations manual. The joint board's separations recommendations are intended to flow through to the individual services that they affect. This must be kept in mind when looking at new pricing methodologies such as price caps. If there is going to be a $2.5 billion cost shifted from the interstate to the state in aggregate, as costs are shifted prices for services in both jurisdictions will need to change. One of the assurances that the FCC gave to the states is that all cost shifts will be reflected in low interstate rates for both AT&T and the LECs. These changes in separations rules are major and will impact new pricing rules that are set up at the federal and state levels.

» *Monitoring*

While the marketplace is awaiting all the new pricing initiatives that the joint boards and the FCC have set into motion, the regulators are concerned with the outcome. To measure the results, the FCC and the joint board have established a monitoring docket in FC CC Docket 87–339. That docket is monitoring eight general categories. One of those categories is pricing of interstate services and state services. How are those services changing over time? What are the benefits to the consumer? The eight areas that are being monitored consist of subscribership, life-line plans, cost and high cost, growth in the network, rates and revenues, bypass, pooling and rate deaveraging, and jurisdictional shifts.

Another area that will affect pricing is with the three life-line plans established by the joint board. The most recent plan, Link-up America, is targeted to improve the household penetration rates. It will reduce the prices for connecting new services to the network and will stimulate network usage. The joint board will examine the plans and determine how or if they improve subscriber penetration. While the HCF is designed to do exactly the same thing, it is also designed to keep rates below one and one-half times the nationwide average cost. These life-line plans are meant to encourage people to participate in the telephone network and will have a major impact on the price for basic local services, as well as for toll and access services.

Rates and Revenues

How are the rates and revenues for both state and interstate services changing, how are they being priced, and what are the benefits and disadvantages of various options that some companies have in place? The joint board hopes to get information in the monitoring docket to address many questions about separations and changes that are occurring in the marketplace, such as: What effect does either economic or uneconomic bypass have? To what extent is bypass occurring?

The joint board concluded in 1987, based on the record, that bypass does not exist. While there are isolated cases of alternative providers of service, or of people putting in their own service, strictly speaking this is not bypass but competition. The joint board, however, wants to obtain more information on the whole issue.

Pooling

The joint board has recommended that the amount of pooling be reduced and limited, which will have a direct impact on pricing. This raises other issues that monitoring hopes to address: What will happen to the companies that stay in the pool? What happens when the companies leave the pool, and how will the resulting deaveraging affect the prices charged to the consumer? Deaveraging is a major issue that must be addressed, in addition to any possible revenue shifts by companies.

Quarterly reports on the eight monitored issues will be published by the joint board staff and will be distributed to Congress and all state regulators.

» Current Issues

The joint board is currently examining a number of issues, one of which is the central office area, as mentioned earlier. It is reviewing dial equipment minutes of use for allocation, versus switched minutes of use of the local dial switch (Category 3). The fundamental difference between the two allocators is the count of local exchange minutes. As DEM is defined, a local exchange intraoffice minute through a switch today is counted as two minutes for every minute. All interoffice calls are counted as one minute. Under an SMOU allocator, local intraoffice minutes through a switch are counted as one minute for every minute, and interoffice calls are counted the same as under the DEM allocator. Whether or not DEM is an appropriate allocator with digital technology is open for question. It makes a major difference in how much will be allocated to the interstate jurisdiction—to the tune of about $300 million. The joint board is trying to eliminate, as much as possible, any administrative costs and burdens that are involved in the separations and allocations process. The joint board is examining studies that have been used in the separations manual over the last thirty years to determine whether they meet today's technology, or if the studies can be simplified to reduce administrative costs.

The joint board also issued a data request to the telephone companies and a Notice for Comments regarding Category 4 of central office circuit equipment and for outside plant cable and wire facilities. The joint board is seeking comments and data on how to further simplify this area. Not at this time but in the future, the joint board will probably revisit COE categories, for the same reasons.

Currently there are four categories of cost, each allocated separately. In the future, there could be further simplification—not only collapsing categories but reducing the types of allocators used to simple, single factors, instead of the multiple factors used today.

All the outside plant cable and wire categories could possibly be simplified, too. Instead of eight or nine different major categories and hundreds of subcategories, this could be simplified into three or four major categories, with simple allocators within these categories.

Another potential concern is host-remote allocation. How should costs be allocated? Today, more and more companies are utilizing host-remote complexes, and the allocations process is not clear. Also, a number of people are questioning whether the current allocation mechanism is possibly wrong.

Today a number of different allocators are allocating different plant, depending on how the company defines the allocators or on how the company negotiates basic arrangements with customers to provide plant. For example, a local loop is allocated to the interstate item at 25 percent plus the HCF. HCF could be significant for some companies. A WATS or private-line loop could be allocated 100 percent to the interstate. It is still the same loop, still the same facility used by the customer in the same way. However, it is priced differently—significantly so. The same local loop, depending on how a customer is served from the local switch to the customer location (through local loop concentrators, through host-remote line units, or through remote line units), makes a significant difference in the amount allocated between the state and interstate jurisdictions. The percentage allocated to interstate varies by company.

Because of the various means in the separations manual of allocating the same equipment, the regulators and some in the industry think that there is a need for change. Some feel that when a customer requests service from the central office to the customer's point of connect, the exchange should be allocated on the same basis, no matter what technology or equipment is used. Obviously, if these changes are implemented they will affect pricing for years to come.

Another issue of concern to regulators is the flow-through of all of those costs. How will the allocations flow through to tariffed services and, ultimately, to the customer who is paying the rates? The concerns regulators have are that with all the shifting of cost from one service to another, are customers getting any benefits in the way of reduced costs of other services that they could not afford before? Is demand being stimulated because of cost shifts? Are all costs and benefits being passed through to all the consumers equally?

These are some of the issues on which the joint board is working. Industry participation in the joint board process is greatly appreciated, especially with respect to the data requests that are issued. The joint board needs to hear from the industry, and the industry needs to express its concerns with the proposals of the board. The most important point, however, is that the industry needs to get involved and be part of the process.

≫ PART VI

Perspective of the Futurists: The Technology Drivers

20 » Network Technology and Information Services: Open Network Architecture

Kenneth R. Raymond

» Introduction

The telecommunications requirements of the customers of the NYNEX Corporation are becoming more sophisticated every day. The network of the future must be sensitive to these evolving needs and must provide customers with the right service at the right price. Open network architecture (ONA) permits the utilization of technological advances; in addition, it provides services that fit present and future customer needs, while enabling competition to affect the marketplace. This paper will give insight into network technology and information services from an ONA perspective. It will also examine the marketing response to ONA opportunities and issues that impact the development of the network using regulatory and product selection criteria.

In June 1986, the Federal Communications Commission (FCC) released its order in the Third Computer Inquiry, which replaced structural separation for the enhanced service operations of AT&T and the Bell operating companies (BOCs) with nonstructural safeguards.

Although the basic/enhanced dichotomy was maintained, the FCC imposed comparably efficient interconnection (CEI) and ONA as the principal conditions for the provision of enhanced services on an unseparated basis. Under this order, if AT&T or a BOC wishes to provide, in the near term, a specific enhanced service on a structurally unseparated basis, it must provide CEI to the underlying basic network for those competitors wishing to offer the same enhanced service. Such basic services must be available to other enhanced service providers (ESPs) in a nondiscriminatory manner.

From a technological perspective, it seemed as though the FCC was describing a totally new network: a network with more distributive capabilities, a network that does not allow a single technology vendor to be the sole provider of a particular type of hardware and/or software, a network that offered substantial freedom for customer applications that were not evident before.

When NYNEX began to formulate its response to the FCC's ONA initiative, it quickly discovered that there was not much that vendors could offer in the time frame set forth by the FCC, and that even if the vendors could provide additional technological capabilities in a timely manner, the BOCs probably did not have sufficient time to deploy enough of the technology to allow ubiquity or standardization. Therefore, what began as an analysis from a technological perspective evolved into an analysis from a marketing perspective: the NYNEX Market Outreach Program was initiated. It sought to inform customers about network capabilities and to respond to customer requests by defining appropriate service offerings. The analysis was then expanded to include services that are needed for the information age but that, as yet, have been neither deployed nor developed.

» NYNEX ONA Plan

On 1 February 1988, NYNEX filed an ONA plan that provides its customers with a robust group of services. Under the NYNEX ONA plan, ESPs will be afforded CEI to the network's basic services, from a technical as well as a pricing perspective. Both customers and ESPs with physical access to the NYNEX telephone companies' networks will have an increased range of features and options from which to choose, allowing ESPs to offer sophisticated new services that take full advantage of NYNEX's network functionality.

The NYNEX commitment to ONA as a long-term regulatory regime is based on the belief that both the public and NYNEX will benefit from an ONA plan that ensures equality among all ESPs and, hence, fosters both competition and the delivery of a myriad of new services to the marketplace. The NYNEX ONA Model, shown in Figure 20–1, provides the technical basis for the ONA plan. To assist in understanding this model, a review of the terms contained therein is appropriate:

- A basic servicing arrangement (BSA) is a combination of network components (communications path, network functions, and network features) essential to the provision of a communications service.
- A basic service element (BSE) is a network function or feature that is optional to a fundamental BSA.

BSAs and BSEs are network capabilities to which an ESP would normally subscribe, or to which an end-user would subscribe for no purpose other than accessing an ESP. In other words, these capabilities are intended for use by ESPs, but customers other than ESPs may also subscribe to them. Also, absent the qualifying statement that BSAs and BSEs are network capabilities to which an end-user would subscribe for no purpose other than accessing an ESP, every network capability used by ordinary subscribers would unnecessarily be treated as a BSA or BSE. Therefore, as implied by the dashed line of Figure 20–1, the end-user's access link (and any other fundamental capabilities with which it is necessarily associated) would or would not be part of a BSA depending on whether it is used for a purpose other than accessing an ESP. Conversely, all BSAs include an ESP access link because it is a fundamental capability; that is, other associated BSA capabilities would have no utility to an ESP without the link necessary to access these capabilities. An ESP access link can therefore be defined as the transmission facilities that connect the ESP to other fundamental capabilities at the ESP's serving central office.

In its ONA plan, NYNEX offers a total of fourteen BSAs, divided into two generic categories: switched and direct connection. The four switched BSAs are distinguished by the type of switching technology employed (that is, circuit or packet switched) and the type of switch termination (line or trunk side) to which an ESP access link interconnects. Various BSE options are available for use

FIGURE 20–1

» *NYNEX ONA Framework Model.*

BSAs
Fundamental combination of service-specific network capabilities, including alternative capabilities that constitute a basic communications service.

BSEs
Network capabilities optional to fundamental BSA

BSAs & BSEs
Network capabilities to which an ESP would normally subscribe, or to which an end-user would subscribe, for no purpose other than accessing an ESP

with each of these BSAs. There are ten direct connection BSAs. Analog direct connection BSAs are differentiated by spectrum and bandwidth capabilities. Digital direct connection BSAs are differentiated by bit rate capabilities. BSE options are available with most of these BSAs. In total, there are ninety-one BSEs underlying and supporting the fourteen BSAs.

» *The NYNEX Network of the Future*

The NYNEX vision of the network of the future expands the principles introduced with ONA and is focused on providing customers with the most technologically up-to-date telecommunications services possible. The telecommunications environment of the future will be driven by a partnership of technology and marketing. In the short term, marketers will specify the features to be included in the services offered to our customers. In the long term, however, technologists will establish a menu of technological possibilities for service offerings, and the marketers will select from that menu the services the public needs and is interested in.

The fabric of the future network will contain emerging network technologies that are on the leading edge of development today. The breakthroughs in technology that will drive the capabilities of the network in the future are occurring rapidly, especially in the areas of photonics, informatics, ergonomics, and artificial intelligence (AI). Through photonics, specifically the wide deployment of fiber-optic cable, it is possible to achieve economies in switching vast amounts of information over great distances in the form of light. Broadband networks of the future will utilize transmission systems based on single-mode fibers and lasers; these systems will offer tremendous amounts of bandwidth to the network, as well as other benefits, such as flexibility, reliability, and security. Informatics, which is the ability to develop software programs that will take advantage of new technologies, is the key to bringing the promises of the information age to the everyday subscriber. Ergonomics involves human interaction with technology or machines. If new services are to be brought to the public, they must be easy for people to use. Ergonomics will help humans communicate with machines just as easily as humans communicate with each other. Humanizing the person-to-machine interface is important because, if the level of difficulty of use of a service is greater than the benefit of the service, that service will not be introduced. AI, another example of informatics technology, is especially noteworthy with respect to telecommunications/information applications. Such AI applications rely on the acquisition of knowledge from human experts. These expert systems are knowledge-based and can assist in the design, development, operations, surveillance, and maintenance of telecommunications systems and the network. Applications of AI could migrate to future service offerings, thus permitting the availability of these benefits directly to customers. These technologies will be integral parts of the NYNEX network of the future.

The network elements that are the prime focus of the NYNEX network of the future are access, transport, and network intelligence. The network is likely to evolve by implementing the new technologies within those elements, thereby creating a modern network fabric more responsive to customer needs.

Network access—the physical connection between the user of our services and the telephone company central office serving that user—will continue to evolve towards electronics-based loop systems using copper and fiber-optic cable technologies. These loop systems will have functionalities that will afford integration and flexibility as to what services are designed and how they are provisioned.

Network transport, the connection between network nodes, is rapidly evolving to digital fiber-optic systems, operating at speeds that range from 45 mega bits per second (Mbps) to ultimately 2 or more giga bits per second (Gbps). This transport capability has traditionally been utilized for the interconnection of network nodes, predominantly through the highly efficient trunk connection. Network transport will feature digital hubs that will support facility reconfiguration and flexibility, thus eliminating much of the physical cross-connection work currently needed to effect customer-driven network changes. This would support a broad range of facility reconfiguration capabilities. In addition, out-of-band signaling transport is needed to connect signaling and control nodes apart from the traditional network trunking. For example, it is anticipated that network management will be linked throughout the network using out-of-band signaling transport. In this way, transport will play an important role in supporting the deployment of additional distributed functionality throughout the network.

Increased network intelligence (feature functionality) will likely provide improvements in switching technology so that current limitations of design and use will be less of a factor in planning and deploying new services. The present generation of digital switching systems being deployed in the NYNEX network goes a long way in providing an expanded base of the most modern feature functionality technically available today. Switching systems that are nonblocking and support distributed as well as improved feature creation in an even more flexible manner are likely in the future. These switches would utilize out-of-band signaling techniques to provide distributed access to new basic network functions, such as new capabilities relating to supervision, status, alerting, addressing, and routing. This type of network architecture has the potential to provide capabilities that have been requested of NYNEX under ONA, but cannot be provided in today's network.

Switching systems would be interconnected to service-specific nodes via out-of-band signaling systems. This type of distributed network control will expand the opportunities to create new and innovative basic network services. The beginning of this network of the future is currently being deployed in the NYNEX network. There is every likelihood that, in just a few short years, this concept of distributed processing within the network infrastructure will make available a broad array of services not thought of at this time.

These technological trends promise great advances in network access, network transport, and the network intelligence of switching and signaling systems. As these technologies lead the way to the NYNEX network of the 1990s, the opportunities for a myriad of new information services being made available to a broad base of Americans through use of the public switched network appears possible. The NYNEX ONA Plan, specifically the ONA Model and the commitment to the ongoing NYNEX Market Outreach Program, will be an important, if not critical, input to planning and implementing network architecture evolution, irrespective of whatever specific technologies are ultimately deployed. This structure provides customers with the ability to obtain access at their premises to features and functions resident in the NYNEX network infrastructure in a timely and appropriate manner. Without a physical access connection to network features and functions, and without the transport and usage interconnection capabilities inherent in the network, customers could derive no utility from the public telephone network. When one considers even the most visionary of predictions of breakthroughs in the areas of electronics, fiber optics, artificial intelligence, ergonomics, and so forth, one finds that they fit well within the NYNEX ONA Model. The evolving architecture promises a continuity of the BSA construct, while making new optional BSEs available. The distributed nature of many of these potential BSEs will enable NYNEX to provide a quicker response to requests for unique functionalities than can be provided in today's network.

» *Summary*

NYNEX views its mission as one of combining the most efficient access, transport, and intelligence elements in a robust and responsive network. The evolution of its ONA plan is linked to the evolution of the actual network architecture. NYNEX is committed both to ONA and to evolving the architectural infrastructure of its network. It will strike a proper balance between technology evolution and the need for network access to the broadest array of competitive information services.

21

» *Network Evolution in the Intelligent Era*
Thomas A. Gajeski

Understanding the current environment is usually critical for comprehending future change. The telecommunications network is no exception.

Today's network supports a wide variety of interfaces using a multiplicity of physical and signaling standards. There are few software interface standards; operations, administration, and maintenance interworking between systems is either difficult or impossible. This environment is a result of the network having evolved around centralized high-capacity switches, with software providing both switching functions and network control functions.

The network has evolved from two distinct components—switching and transmission. The distinction between these components has become blurred with the implementation of digital technology. Initially, digital carriers systems were placed on existing copper pairs to increase capacity with low additional capital cost. To take further advantage of these transmission facilities, digital switches quickly made their way into the network. With these switches came a distributed architecture whereby switching matrixes were placed remote to the central office and closer to the customer.

Concurrent with the merging of switching and transmission facilities is the change in system design from large centralized processors to distributed processing. Key characteristics of these new switching systems are modular design, clearly defined functional responsiblity, and a flexible topology. The introduction of digital-switching distributed processing is the key to the current evolution of network capabilities. Advantages of this new distributed environment are flexibility, reliability, and capacity.

These changes in telecommunications equipment provide the opportunity to use this technology for the evolution of the structure of call handling. The resulting equipment design has provided the local exchange carrier (LEC) industry with the opportunity to modify the basic network architecture. These changes are driven by the need to transmit more information as part of the evolving competitive environment.

The architectural goal is to separate current signaling and control functions from the transport function. Implementation of this approach will require that equipment be designed with standardized interfaces, modular functions, and an open architecture structure. The LEC must be able to add different vendors' equipment to existing systems without limiting new equipment purchases to vendors of existing equipment.

With the increasing separation of control from transport, the network will become an intelligent network with the ability to provide new services without major additions or changeouts. In addition, all network elements still have access to control and data base functions. This will provide a challenge to properly price new services.

The desirable future network architecture will not just happen. It will take careful planning and many evolutionary steps to implement. As with all changes to the network, this will be an incremental evolution that uses today's network. New switches, enhanced processors, network data bases, and high-capacity fiber-optics systems will be added to the existing network. The first step is currently underway, that being the creation of a separate signaling network.

The implementation of a common channel signaling network using signaling system No. 7 (CCS/SS7) has already started and will be the cornerstone of further network evolution. Currently, the network can be displayed as end-offices connected to access tandems, which in turn are connected to interexchange carriers (ICs) (see Figure 21–1).

FIGURE 21-1

» *Existing U.S. Telecommunications Network.*

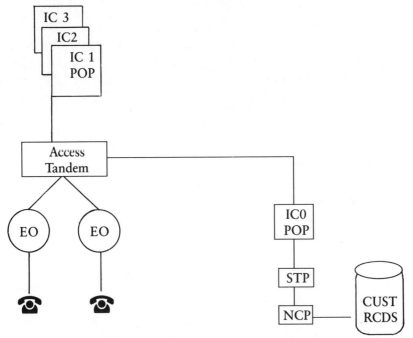

Terms: interchange carrier (IC); point of presence (POP); end office (EO); signaling transfer point (STP); network control point (NCP); and customer records (CUST RCDS).

Overlaying the CCS/SS7 network on the existing network will require creating interfaces between the existing switches and this new network. To better understand how these changes will occur, an explanation of the components for the new network is required. The first components of the CCS/SS7 network are the signaling point (SP), which is the interface between the switching office and the CCS network (see Figure 21-2). The SP may be a switching office, a data base, or any other signaling node. The switching office function is to convert interfaces such as dual tone multi-frequency (DTMF) to the SS7 protocol. Eventually, and depending upon local service requirements, the SP will be deployed within the local end-office.

The service switching point (SSP) is a software feature capability within a CCS switching office for such services as 800 data base service and private virtual network (PVN). The SSP is a stored

FIGURE 21–2

≫ *CCS/SS7 Network Overlay.*

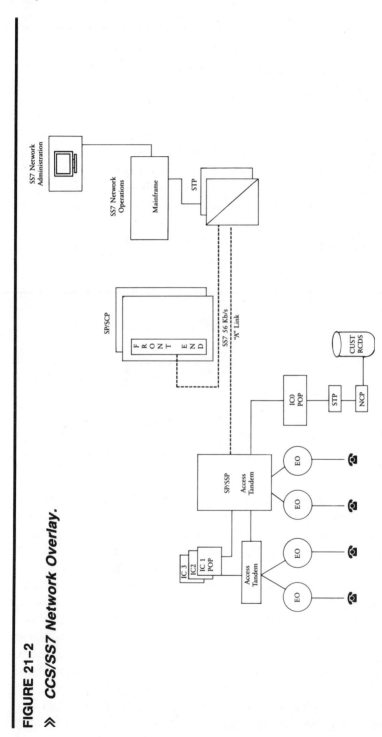

Terms: interchange carrier (IC); point of presence (POP); end office (EO); signaling point (SP); signaling transfer point (STP); network control point (NCP); customer records (CUST RCDS); service control point (SCP); kilo bits per second (Kb/s); and service switching point (SSP).

program–controlled switch equipped for SS7 capability and is a specialized signal point that provides CCS network access. This is done by loading the appropriate parts of the SS7 protocol for the services handled by the particular switch.

The signaling transfer point (STP) is a packet switch that routes and concentrates SS7 messages from switching point to switching point. Incoming messages are directed to the appropriate destinations depending upon the header address information of the packets. The STP receives messages from an SP, directs them to the appropriate service control points (SCPs), and delivers the retrieved or processed information to the SP and/or other SPs, or perhaps to other data bases.

The SCP controls access to data bases used for various functions such as number translations, carrier identification, standard trunk selection, and credit card validations. The maintenance and updating of the data bases associated with SCP is from both local and distant centers. For example, the STP or 800 services will be administered and updated by the national service management system (SMS) in Kansas City.

Signaling links provide access to all nodes (SP, SSP, STP, SCP) on the CCS network via 56/64 kilo bits per second (Kbps) digital data links. These links are grouped into categories according to the nodes they interconnect. Access links (A-links) connect SP/SSP switches to STP pairs and are provisioned in quads. Connecting links (C-links) connect mated STPs. Diagonal links (D-links) connect STPs of different levels (local versus regional) and are provisioned in quads (see Figure 21–3).

The SMS is an operations support system that provides service creation and customer control capabilities, as well as support for the administration, coordination, and control of data bases for a particular service (for example, 800 service). The primary function of the SMS is to administer the customer call processing records in the SCP data bases. SMS handles call processing record additions, changes, and deletions by dial service administration centers (DSACs) or allows changes to be made by individual subscribers. The SMS takes this input, validates it, translates it into the format required by the SCPs, and distributes it to the appropriate SCP data bases. The service administration center (SAC) is the support center for the SMS.

The separate signaling/control network outlined above will support the addition of new service with minor changes to the basic

FIGURE 21-3
» *Operating Overlay.*

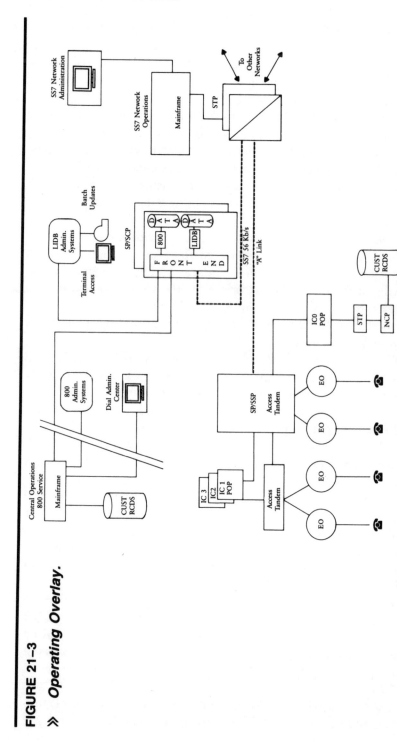

Terms: customer records (CUST RCDS); line information data base (LIDB); signaling point (SP); service control point (SCP); kilo bits per second (Kb/s); signaling transfer point (STP); service switching point (SSP); interexchange carrier (IC); end office (EO); point of presence (POP); and network control point (NCP).

structure. While the following is not a complete list of such services, it does include those that will be among the first offered with the use of this new network.

Alternate billing service and exchange access operator services will require SS7 access to obtain calling card validation and IC routing information. This service will be installed on the switching systems deployed for operator services when the regional Bell operating companies (RBOCs) vacate the current operator systems shared with AT&T. Calling card information from the customer originating a call is sent to a data base containing the customer's subscription information for validation and determination of the designated carrier.

A data base used to direct calls and identify the address of the caller will improve 911 service. It will also provide instructions and caller-specific information, such as previously submitted medical data.

Custom local area signaling service (CLASS) is the first service that will require the deployment of SS7 at the end-office level of the network. This service transports the calling party identity to the terminating switch via the CCS network. The party receiving the call can implement various screening capabilities on the delivered calling party number in conjunction with existing custom calling features. Automatic callback and recall will also be available. In addition, calling number display and customer-originated trace are optional features.

Area-wide Centrex (AWC) permits the sharing of certain features and a common dialing plan for multilocation Centrex customers. This service is expected to be available in 1990 or 1991.

Area-wide message desk service for Centrex customers provides the delivery of calling party information to an attendant or designated station. This feature greatly increases the capability of the message desk service currently available and will be deployed in conjunction with AWC. Deployment in a calling area with CLASS will greatly enhance this feature, as it will simplify trunking to the message desk by eliminating trunks to each served office.

ISDN (integrated services digital network) is a network architecture that will ultimately be capable of supporting existing intelligent network (IN) services. SS7 will be required to expand stand-alone IN/1 and IN/2 or individual ISDN switching systems—referred to as islands—into a true configuration for calls between sophisticated terminals. Networks of ISDN are expected to be required in late 1989 or early 1990 as customer requirements demand connectivity to new services and features, as well as simultaneous transmission of both voice and data.

The evolution of technology and the new competitive environment following divestiture have caused many to reassess existing structures for providing service. Some of the resulting changes will alter accepted procedures. Moreover, customers will notice these operational changes before they will notice the expanded network capabilities.

The future changes are not confined to technology. Telephone numbers are a finite asset of the communications industry, and all segments of the industry have an obligation to use this resource in an efficient and conservative manner. Many telephone companies have a growing concern about the effects that new services will have on a dwindling supply of numbers. The administrator of the North American numbering plan (NANP)—Bellcore—estimates that, under its present structure, NANP will exhaust its supply of area codes by 1995. An acceleration of this date will prove extremely costly to the industry. Many new services require new numbers. For example, a full dedicated prefix within the geographic NPA may be used to access a new service (for example, 976). The method based on NANP then uses normal three-digit routing and allows the collection of distinct billing information. Although this option provides a workable approach for new services' number plans, it does shorten the life of the current NANP.

Current discussions in the industry would identify ISDN networks by an "area code" that will be independent of geography. Just consider what typically happens when you receive a telephone slip stating that someone called while you were away. If you do not know the current location of the individual, you look at the area code to determine his or her location. When to return the call may depend on your respective time zones. With the new ISDN numbering proposal, this will not be possible.

The key to creating successful new services is the development of a flexible network. This requires fundamental planning based on the reality of the future telecommunications environment in the United States and the already well-established directions now being taken by the industry. The forthcoming information age will demand many new communications services, but most such services will not be planned. They will evolve from the basic CCS/SS7 capabilities.

In preparing for the future telecommunications environment, it is futile to speculate on the prospects of a long list of individual features. In such an environment, it matters not whether individual features are successes or failures. If failures, they will just go away—but as long as the basic CCS/SS7 capabilities are in place, other features will come along to take their places. And they will do so at little additional cost to the telephone companies because these new features will use the basic capabilities already in place. From philosophy underlying future network planning will evolve the bulk of new services and additional revenues in the future.

22

» Models of Network Infrastructure: Pricing ISDN for Access

Loretta Anania
Richard Jay Solomon

Implementation of an open network architecture (ONA) regime and integrated services digital network (ISDN) technology implies a reallocation of resources among public and private networks. This new telecommunications system is sufficiently different from the past network that transition to a digital infrastructure may be neither smooth nor equitable. An end-to-end digital system will ultimately emerge, with virtual, logical connections defining communications channels. Whether this network will be truly open depends upon the success of nonproprietary standardization efforts.

The allocation of fixed costs in the local distribution network will change in response to technological innovation, new capital and depreciation requirements, and changing regulatory regimes. The question will then be, who pays for this innovation, and how? Choices must be made, and the argument is presented that some of the economic models that were found to work well in the past but are unsuitable now, paradoxically, may work again.[1]

It will be argued that, as telephone networks become complex computer networks, user control—not carrier control—increases. With the user taking control of routing, bandwidth allocation, and administration, there will be unexpected applications of the public network, including rate arbitrage on an international scale! Such dynamic allocation of network resources will become increasingly difficult for the carrier (or regulator) to track. So, with integrated digital networks, the flat-rate, or pay-in-advance subscription solution, may be the best method of pricing.

» Infrastructure

The integrated digital communications network of the future will not look anything like the telephone network of the past, neither in architecture nor functionality. Its usage will be different, more dynamic, and less predictable. To maximize system efficiency, customers, including other networks, will need to gain direct access to network resources, such as operations, administration, and maintenance. And to maximize the profitability and use of the existing plant, carriers will have strong incentives to expand into new areas of business, often in competition with their customers. End-to-end, high-speed digital communication will generate totally different demand profiles. For example, continuous channel occupancy may not be needed for most services in the future. Even movies, properly encoded, can be sent over medium-bandwidth lines in compressed time. Packetized voice needs hardly any bandwidth at all (Solomon 1987).

Changes in technology and social policy are already causing stress to preestablished costing and regulatory pricing practices. Bypass, resale, and access charges to the network exemplify this readjustment. Radical, sudden shifts in regulation, customer demands for network control, and the competitive push for new plant investment may even threaten the major carriers' viability.

To maintain universal access while at the same time expanding network provision requires innovative ratemaking and novel depreciation schemes. So far, public inertia, a regulatory lag, and defaulting to yesterday's tariff formulas have artificially constrained current technological reality, making the economic transition even more painful than it need be.

» *Models*

To see how tariffing will change, it is useful to compare some old and new models for telecommunications infrastructure.

The old system of analog telephony was patterned on a nineteenth-century intercity rail transport model.[2] This telephone model worked out both architecturally (trunk and branch/loop), politically (essentially one carrier and one service, except at major nodes), and for tariffing purposes (distance and time-sensitive variable costs for interregional links). Whether rational or not, it is common to find social regulation of a new technology patterned after some older technology that it appears to resemble (at least at first) (Pool 1983).

Based on this model, a system of telephony in the United States developed to meet the following policy objectives:

- Universal service, but with businesses and residences splitting the cost of the fixed plant according to a pricing scheme with differential pricing for essentially the same service. Businesses paid value of service, while residential prices reflected ability to pay (even if below cost).
- Capital for the entire end-to-end system was raised by the "dominant" carrier.
- Subscriptions were for a lifetime. The system was on a cash basis, and demand predictions could be based principally on population growth and mobility. Everyone had to pay in advance for network access.
- Cost-engineering function was averaged for the network as a whole, but the network was artificially separated into state and federal domains, which reflected the fact that calls were predominantly local.
- Rates were based on plant investment. With most calls being local, unlimited flat rates for local service stabilized revenue projections.
- No cream-skimming: to prevent erosion of the rate base, no resale and no private attachments were permitted.

Though traffic patterns in the United States changed over the decades, especially with the postwar suburban explosion, this railroad transport model was, until recently, rarely challenged. Beginning in the 1960s, the telephone network underwent a fundamental transformation through the incorporation of digital connections and signal processing. Adoption of digital computer technology was essential for modernization and economic efficiency.

The current network is computer stored program controlled (SPC) switching, provides different levels of network access, and connects sophisticated customer-owned equipment, including other networks. As new services other than POTS (plain old telephone service) abound, at least in name and in advertising vaporware, the available rate structures vary tremendously. They range all the way from life-line flat rates to complex Centrex packages.

This network is very different from the original model. The assumptions of the old integrated national network included: fixed overhead, fixed bandwidth, physical analog connections, and a hierarchical network architecture. With new public and private networks, based on the digital computer, a new set of assumptions is required: variable bandwidth allocations with logical, instead of physical, connections and a nonhierarchical network architecture under shared carrier/customer control with distributed processing nodes.

On the old network it was possible to separate embedded, subscriber-access plant from interoffice and interregional plant. That distinction became more artificial as more integrated equipment evolved. Still, the economics of regulation and the corporate structures of the telephone companies encouraged this artificial separation between local and long-distance service and the separation of service between states—even when not in the public interest and probably not in the companies' economic interest. In future, such separations just for accounting purposes on all-digital networks will be almost impossible to justify rationally.

» Processing

Voice telephony was intended for everyone, at a cost that local residents and businesses could afford. Subscription was for life, payable one month in advance. Truly differentiated service (as contrasted to differentiated pricing) was difficult and too expensive to provide (and bill for) in any but extreme cases. Then came data processing (DP)—a special, new, and revenue-generating service that not everyone would buy. But demand for DP has grown at a faster pace than the demand for POTS.

Computing has been unregulated, and telecommunications heavily regulated. The two businesses are becoming integrated from a technical viewpoint. Regulatory attempts—Computer I, II, and III—tried (and failed) to resolve the inherent anomalies of telephones that now compute and computers that communicate.

Deregulation is easier to pull off than effective re-regulation! The Computer Inquiries were for the regulatory convenience of keeping DP and telecommunications separate, at least as far as the industry's accounting and revenue base is concerned. But this creates even more economic and political distortions. Defining what constitutes a hybrid service, or an enhanced one, and excising unnecessary protocol conversion (the carriers would not call it processing because that is what computers are supposed to do) defy robustness and are exercises in magic. The derived regulatory separations can only be short-term. Neither regulators nor carriers wish to admit, on the record, that end-to-end digital telecommunications carriage is itself a computer process, using programmable machines that communicate.

Computers do many things, virtually simultaneously, by doing only one thing—adding fast. Simple but profound, this is the magic of Boolean logic, the mathematics of combinations, and high-speed (opto)electronics. A digital switch is simply a fast adder. Processing is an ordered series of instructions for a machine that can follow instructions. A general purpose computer is a fast-adding processor. All digital telephone switches are general purpose computers, and therefore data processors. Switching bits is what a computer does.

There must be a big conspiracy here. Data processing sounds like it means the alteration of information, and this is not what we want governments to regulate; for if it did, it could do nasty things to or with the information, and perhaps to information providers as well. So society pretends that regulators are not regulating information, by calling it something else. If one appropriately defines information transfer as just an electrical process, it is not necessarily the case that data processing will imply content alteration, and perhaps, then, one could admit that digitally switched telephone common carriers are already DP providers. With the network doing de facto processing, pricing becomes much more difficult than with mere transport, as in the original telephone model (based on an analogy to physical transport).

» *Pricing and Capital Formation*

At a minimum, the expansion of the market for telecommunications and information services brings a demand for network digitization and integration to provide faster connections, end-to-end connectivity, and more network control. The trends toward the network of the future include (1) faster and faster computer processors acting as switchers, (2) virtual end-to-end connectivity for user-to-machine, machine-to-machine, and user-to-user communication, and (3) immense network control power in customer premise equipment (which tends to be digital processors, even for voice). All of this leads to flexibility of choice among technologies, carriers, rates, and services.

According to classical pricing theory, in the simplest case of a multitude of entities providing an indistinguishable basic service, price is driven to equal marginal cost. If prices are lower than marginal cost, customers tend to buy more of the service. Even though volume increases, if price does not cover the cost of incremental plant, a firm (or service provider) loses money on every sale.

A monopoly provider (meaning there is no other way to get the service) could choose to charge a price higher than marginal cost. If its service is truly inelastic and we must have its telephones, water, oil, and so forth, at any price, then the monopoly (or cartel) gets very rich indeed: "monopoly rent," it is called. But if the public can do without the service (even though it would rather not), or cannot pay and must do without, then public needs will go unmet. Profit maximization (assuming relative inelasticity of demand) may fit the goals of a private monopoly, but hardly those of a public service operation. In economic terms, marginal-cost pricing tends to maximize social goals, such as universal service for telephony. This is the kind of pricing that local regulators like.

Telephone and telegraph services may have begun as natural monopolies, but regulation and the threat of competition, each in its own way, have prevented very much in the way of monopoly rents. Yet the question remains: How was service to be extended to all those willing to pay at least marginal costs if telecommunications was to have indistinguishable service characteristics?

Pure marginal-cost pricing for an expanding utility produces a deficit that can be made up in several ways: from subsidies or special assessments on users to pay for modernization; or—catastrophically—permitting an obsolescent entity to go bankrupt and then nationalizing the firm at bargain values (or if it is a bankrupt state entity, denationalization at bargain values); or some other form of corporate reorganization. Cross-subsidy, according to the classical theory, would be out of the question, except perhaps as direct cash flows to the indigent user. And capital formation for expansion without sufficient profits would be even more difficult (indeed, it has been for nationalized firms).

But there is a less catastrophic way. Deficits can be filled by disaggregating services: finding some services with increasing costs to be priced above marginal cost for different markets, such as PBXs and key sets for businesses, POTS for residences. Monopoly privilege helps enforce such disaggregation.

One basic factor in marginal-cost pricing is that telecommunications, so far, has been a declining cost industry in which marginal costs drop as production is expanded. Rapid technological change in telecommunications, even without expansion, appears to reduce marginal costs even more than average costs. If marginal costs are below-average costs, and price is set to marginal cost, total revenues will still be less than total costs.

Despite rapidly declining marginal costs due to technological advances, the telecommunications link traditionally has not decreased in price as fast as advances in electronics have lowered computing costs and increased computer processing speeds. In the future, however, with high-speed fiber, telecommunications throughput may overtake most computer bus (input/output) speeds. For the first time, network resources may be faster than the terminal equipment can support.

This is an extremely significant change. It may reverse the historical trends of computing and communications costs. Not only will costs reverse, but demand profiles will be greatly different from today's. For example, demand for fiber-based telecommunications may come from fiber-based, local area network users.

Altered demand profiles, no matter how potentially profitable, are a mixed blessing for carriers that have been gearing up for a different market. In today's rapidly expanding and changing computer-communications plant, there may not be enough money for tomorrow (Anania and Solomon 1987b; Leisner, et al. 1987). For example, the telephone industry's internal reserve ratios have consistently declined since World War II. In 1980, the ratio bottomed out at 18 percent, whereas in 1926 it was as high as 29 percent.

A decline in internal funds, plus increased competition for outside sources of capital, without someone to foot the burden of increased depreciation costs will substantially limit the telephone industry's ability to replace and modernize its plant (Egan 1987).[3]

Capital recovery for modernization is a more critical issue in telecommunications today than is the question of whether basic or enhanced services should be offered, or what color and label the cheaper phones should have.

Coming from the (as yet) unregulated part of the communications business, the computer community has been proceeding with its own independent research on protocols, data base technology, and networks, outside of the mainstream of international telecommunications standards forums. Their solutions, based on DP techniques for very fast data and voice communications, may make the original ISDN standards and concomitant open systems interconnection (OSI) framework obsolete before they are even fully implemented.

Interestingly, broadband for computers does not mean continuous holding times for a circuit-based wideband link. Instead, demand is more like that on an intermittent (asynchronous), wideband computer bus, extended beyond the computer's own confines. In other words, a wide-area extension of a local area network.

Television can also be asynchronous. Video information is transmitted only when sections of an image change. There are other possibilities, such as sending raw information needed to recreate a full movie in compressed time—that is to say, transmitting the data faster than the speed of ultimate playback, called "burst" mode.

A further discussion of demand for broadband is outside the scope of this paper, but these intermittent broadband applications have drastic implications for future telecommunications network demand profiles.

» *Broadband in the Loop*

An irreversible trend toward computer-based, customer-provided terminal and even switching equipment, enhanced by relatively easy connection to an all-digital virtual network, shifts a significant portion of what formerly would have been carrier costs to the user (Anania and Solomon 1987a). Virtual networking entails network transparency for optical broadband interfaces. Just how these interfaces are to be chosen is currently the subject of much debate on the international scene.

Broadband ISDN (B-ISDN) concepts are an order of magnitude

more powerful than narrowband ISDN (N-ISDN), with potentially a steeper decline in transmission and switching marginal costs. If wire cannot be easily upgraded, single-mode fiber can. This fact, along with the possibility of altogether new architectures for super-fast switching, led originally to an effort to plan for a smooth evolution from narrowband to broadband. However, it may turn out that at little extra cost, a B-ISDN network will leapfrog N-ISDN technology.[4]

ISDN standards committee meetings have been operating for almost a decade paying little attention to distributed microcomputer technology. Now that computer engineers have taken a more active role in international telecommunications network standardization, with novel software-defined architectures, the development of fast-packet switching, and especially single-mode fiber, B-ISDN is becoming altogether different from N-ISDN.

N-ISDN is an end-to-end digital solution offering a basic rate of 160 kilobits per second (Kbps) or a primary rate of 1.5–2 megabits per second (Mbps). These rates were primarily intended for existing twisted copper–pair plant, though the primary rate uses mostly coaxial cable or fiber today.

B-ISDN is different, though it subsumes the narrowband "2B+D" channel rates. B-ISDN standards are expected to apply sophisticated, relational–data base software across a high-speed digital link for both transmission and routing, rather than using simple bit interleaving (as used in T1 telephone carriers). This gives the switch enormous power, but it is under the control of the subscriber, since the packets are essentially self-routing in most B-ISDN proposals.

Furthermore, broadband fast switching and cross-connect technology will demand a return to inband signaling. But "inband" here is quite different from Consultative Committee on International Telegraph and Telephone (CCITT, 1987) No. 6 tone signaling.

The current standards call for an H_4 rate—approximately 150 Mbps. (Rates of more than one gigabit per second may also be introduced in the near future.) The key to this network's design is the use of enormous bandwidths for overhead; such bandwidth is available for the first time with single-mode fiber. Broadband fiber payload is ninety times larger than the primary rates, or 1,000 times the basic rate for a local loop. Indeed, overhead alone is about ten times larger than the entire payload capacity of primary circuits.

Most important, for the H_4 interface arrangement, unlike one-dimensional ISDN and primary rate framing, the "packet" is in the form of a matrix, replete with pointer cells. It is these pointers that allocate the system resources. This matrix can accommodate various headers and subheaders in an envelope. This is what makes these very high-speed packets or envelopes of frames self-routing (independent of nodal hierarchies).

Envelopes permitting variable bandwidths give users powerful network control options without reducing the carriers' management and operations control. The new software-controlled options could be designed to permit carriers and customers to share control of a "virtual" network, without mandatory colocation. This is what the Federal Communications Commission's (FCC's) ONA proceedings should be all about.

An H_4 channel could be equal to about two digitized, compressed, but *continuous* high-definition TV (HDTV) signals, or 4,500 uncompressed, simultaneous telephone calls. But, as was noted, sending such data in real time is only one way to use B-ISDN standards. The beauty of the frame matrix structure is its elegance in handling multiple services and bandwidths in very short time intervals (125 microseconds per frame) and in facilitating video in delayed time, data in compressed time, or packetized voice. Again, it is emphasized that B-ISDN does not necessarily mean wideband, continuous occupancy of a channel.

These drastic ratio differences between N-ISDN (and analog telephony) and B-ISDN are bound to affect tariffing theories and increase the range of user-defined services. Moreover, as long as carriers have to install single-mode fiber for the distribution network and are willing to adopt B-ISDN frame interfaces, it makes little difference in cost whether the packets are running at 151 (Mbps) or one gigabit per second!

» *Arbitrage*

Integrated digitization of the network will not permit temporary cross-subsidization and price discrimination among customers who have the know-how and resources to bypass. The resources may be economic, technological, or political. Because control of the bitstream is shared between the carrier and the customer (and the customer may be another network), the regulatory distinction between user and carrier becomes as blurred as the separation between services.

When further evolution towards the implementation of B-ISDN variable-rate networks proceeds, network control may become the most important commodity being bought and sold in the telecommunications market.

Under these circumstances, how is it possible to justify different network pricing for data than for voice when these are physically indistinguishable bits on a fast link? Continuing to charge differential prices only leads to inefficient use and network arbitrage.

Arbitrage occurs when there is a discrepancy between price and cost, yielding an opportunity for a third party to profit by reselling. In some circumstances, this is not quite legal, but there may be ways around it. Arbitrage is a market concept that has not been seen before in broadband telecommunications; but, as will be demonstrated, with deregulation and virtual end-to-end digital networks, arbitrage is becoming increasingly viable as long as carriers maintain a significant differential between actual costs and charges to the customer.

The new technology of integrated digital systems could result in a distinct paradox when it comes to tomorrow's carrier revenue. Though it may appear on the surface that putting all telecommunications services onto one network enhances the concept of natural monopoly (and monopoly rents for captive users), the integration of these services into just one form of carriage in an invisible and nondistinguishable digital bitstream has created economic pressures for cost-based pricing.

How can this revenue paradox be? ISDN promoters claim that the technology will permit the abolition of such things as private lines and other forms of competition to carrier monopoly. The answer lies in simple economic common sense: in the long run, an educated customer cannot be charged more for one service when there is a cheaper substitute service available (albeit under a different name).

Some carriers currently view ISDN as a way of offering value-added DP services, protocol conversion, and information manipulation. They see this as a new way of establishing new multipart tariffs based on differential value of service. Where business uses predominate, service "kinks" like call forwarding and camp-on are priced at a higher level than the marginal and average costs of their digital software. Where traffic-sensitive rates apply, average costs should then replace marginally priced flat rates for private lines and local calling.

There is a flaw, however. The new technology brings more than tariff and price changes, it brings incentives for usage change. With

ISDN, or any all-digital switched network, subscribers have the option of paying only for pure bit transmission and providing for most enhancements with their own resources, wherever arbitrage makes it worthwhile.

The customer may be given some opportunity to whipsaw the carrier for a change. The technological know-how includes use of N-ISDN's inherent protocols, such as CCITT Signaling System No. 7 (SS7), and its extension on the D channel has a peculiar technical loophole: the first segment of its binary octets (bytes) intended to signal user-to-user (UTU) information can also be used to send data from the caller to the callee before the call itself begins. (This has engendered a controversy over whether all call attempts, as such, should be charged, even if not completed.)[5] With such precall data links, there is little in enhanced services that a carrier can offer that a sophisticated user cannot get from well-programmed, customer-premise equipment (CPE). Computer-based CPE technology, therefore, will drive carrier pricing close to marginal costs.

With B-ISDN envelopes (domains) and pointers working like tuples and attributes, or rows and columns in a data base, even more powerful customer control can be imagined. Oddly, if carriers attempt to forbid use of the signaling protocols this way, neither N-ISDN nor B-ISDN can work. For example, unless the CPE can handle protocols before communication begins, the very premise of ISDN having a small set of uniform terminals will be violated, and whatever the service is, it will not be ISDN. Similarly, with broadband interfaces following open architecture principles, part of the frame overhead must be customer-accessible. If carriers restrict use by monitoring content and terminal type, this would likely drive the largest users to build their own networks that will offer flexible, dynamic, frame-structured wideband services.

» Bypass

Instant Traffic Diversion

A basic ISDN principle being promulgated, though not yet adopted by the CCITT, is that an ISDN voice call between two points will not cost any more than a voice call over the conventional public-switched telephone network (PSTN) between the same two points. The FCC proceeding on access charges for enhanced carriers (nonvoice is all the

FCC really means) was intended to redistribute perceived, uneven local loop costs between voice and data users. Similar future attempts to charge more for a data call than a voice call on an integrated digital network will be impossible to enforce, since the bitstream will be indistinguishable by either the carrier or the regulator.

Assuming voice is cheaper than data, equal charging will make it possible to seize a 64-kilobit digital line—at the bargain-basement voice rates. If the carrier needs to know, the customer tells the carrier (digitally, of course) that it is talking on its own network—but in reality it subdivides the circuit into two, four, six, or more voice/data circuits, using sophisticated voice and data compression equipment already on the shelf. The carrier will be charging for only one voice circuit, and the customer can hang onto it all day, creating a virtual, multicircuit tie line.

Leaky Digital PBXs

ISDN CPE will enable you to redistribute (switch) these subdivided channels locally. Writing software that can establish such connections transparently will be a growth industry, particularly if such data/voice restrictions prevail (as may be the case overseas).

On-line Rate Shopping

With effective control over route selection shared between the user and the carrier on an ISDN, a virtual network can be designed to take advantage of the best routes and rates dynamically. (ISDN recursively permits bypassing itself.) In nonintegrated systems, it is difficult for customers to do their own voice switching. With ISDN and fast user-premises switching for voice and data, rerouting will be as easy as store-and-forward message drops are now. The least expensive, most reliable route may undercut attempts at revenue requirement stabilization.

An Example Today for the International Traveler

Today, with AT&T and British Telecom calling cards, a regular trans-Atlantic commuter can choose which tariff to use for billing, depending on the current exchange rate. The carriers, of course, cannot change their tariffs as fast as the 24-hour financial markets change currency rates. ISDN is not even needed—just direct, international credit card dialing. (Carrying this logic to its end, eventually a universal currency must emerge based on electrons or bits. With USA Direct or UK Direct, it is well on its way today.) Similar situations will be found wherever a billing arrangement is unbalanced, even in the local exchange, as long as the tariff is based on something other than simple access.

Dynamic Callback

Though carriers may attempt to equalize the accounting rates they charge each other to share a circuit, there may be little correlation between these costs and the prices charged users. Special rates such as WATS and "metro-calling" are examples. An on-line data base, operated by a third party, may tell the sophisticated customer which is the best rate at the moment, and in what direction, and then make the connection.

For all the reasons above, the rate elements that were used to price services on the local loop will no longer be satisfactory. Rate elements based on (1) call frequency, (2) duration, (3) distance, and (4) time of day assume that costs of switching and trunking plant (interoffice and tandem) are variable. Features and functions like call forwarding, speed dialing, and call waiting added a new variable element based on SPC memory and processing increments. But ISDN and B-ISDN technologies are not straight-line extrapolations of SPC circuit switching, especially in the packet mode (for voice and data). Memory tables are inherent in any case for D-channel and SS7 signaling. Adding features is more a matter of toggling bit positions in already existing memory space. Memory space enhancement, therefore, becomes virtual and depends on UTU signaling under user control. In B-ISDN, as already mentioned, user signaling is even more powerful, including routing and temporal feedback information for possible operation, administration, and maintenance purposes.

» *The Universal Flat Rate*

As the separation of basic service elements (BSEs) increases in complexity, the only feasible solution from a user perspective becomes a universal, flat-rate ISDN subscription. Due to the vast capacity of fiber and the switching fabrics under study, it is likely that with B-ISDN fixed costs will overwhelm any variable costs. The variable costs, if any, may be too small to measure or too complex to detect. The basic variable cost often cited—time-of-day traffic, or designing for the peak hour—may find a solution in the asynchronous frame transfers working their way very rapidly through a mesh, instead of a hierarchical network. This means that adjacent exchange areas (all the way to the coasts!) may be relevant to any local tariff scheme. Local will be inseparable from medium distance, if not long distance, in B-ISDN.

Hence, access to the network, based on the maximum bandwidth a subscriber wants on the local loop, may be the only way to tariff broadband fiber. It is likely that even with N-ISDN we will find that *access* is a more productive concept to use to construct tariffs than bits per second.

The sum of customer access costs, then, would be equal to the total revenue requirement for each carrier. This might solve the arbitrage problem and the carrier revenue paradox, to boot.

» Paradigms

With ISDN, narrow or broadband, one has a paradigm change: by permitting multiple and fungible services to be combined transparently on the network, carriers lose the ability to differentiate between services in order to continue cross-subsidization policies. (Nondifferentiation does not help the private carriers either—how profitable would competitive automobile companies be if one car, painted black, satisfied everyone's tastes and needs?)

In a distributed data processing environment, which is what ISDN is all about, the old games to help meet a carrier's revenue requirement will no longer work, for the underlying system design itself subverts the rules. Totally new rules, quite unlike what we have been used to in public utility economics, will ultimately have to be worked out for ISDN or B-ISDN tariffs. During the transition, when the old hardware is being amortized at the same time the new system is capitalized, some bizarre rates and service offerings could appear. Some very painful steps may well have to be faced by the dominant and the competitive carriers to avoid severe revenue losses.

» References

Anania, Loretta, and Richard J. Solomon. 1986. "Changing the Nature of Telecommunications Networks." *Intermedia* (May): 30–33.

Anania, Loretta, and Richard J. Solomon. 1987a. "The Ghost in the Machine: A Natural Monopoly in Broadband Timesharing?" *Telecommunications,* Dedham, Mass. (October): 16–18.

———. 1987b. "Capital Formation and Broadband Planning: Can We Get There from Here?" *Telecommunications,* Dedham, Mass. (November): 20–21.

Consultative Committee on International Telegraph and Telephone (CCITT), Study Group XVIII. 1987. "Transmission Aspects of Digital Networks." Report of Working Party XVIII/7, COM XVIII-R 44 (A, B, and C) (August).

———. 1984. *Red Book* (I-series recommendations). Geneva, Switzerland: International Telecommunication Union.

Egan, Bruce. 1987. "Costing and Pricing the Network of the Future." Paper presented at the International Switching Symposium, Phoenix, Arizona (March).

Leisner, Susan; Jeffrey Rohlff; Chuck Jackson; and Harry M. Shooshan. 1987. "The Effects of Tax Reform on the Telephone Industry." Paper presented at the Airlie House Conference on Telecommunications Policy, Airlie, Virginia.

Pool, Ithiel. 1983. *Technologies of Freedom.* Cambridge, Mass.: Harvard University Press.

Solomon, Richard J. 1987. "Open Network Architectures and Broadband ISDN: The Joker in the Regulatory Deck." In *ICC-ISDN 87—Evolving to ISDN in North America.* Dallas, Texas: International Council on Computer Communications.

» Notes

1. The views and opinions in this article are the authors' only and do not necessarily represent those of their affiliations.
2. Luckily, the telephone analogy was to an intercity rather than a local-transit (streetcar) model, which in most cities consisted of multiple companies with multiple fares and no interconnections, either local or intercity.
3. One solution, sending television to the home over an integrated-fiber net, may have tremendous potential for recapitalizing the subscriber-based network—and solves the spectrum problem, to boot. But this option may be the least palatable, given today's layered, hierarchical, and embedded economic and political obstacles!
4. For a more complete description of the issues on B-ISDN and a detailed technical bibliography, see Solomon (1987). The October

1987 issue of *IEEE Selected Issues in Telecommunications* is ·devoted to broadband switching concepts.

5. For more detailed information in ISDN specifications, see CCITT (1987) and updated reports in this series of memoranda from the CCITT. The complete specifications are being published by the International Telecommunication Union in the CCITT "Blue Book" series during 1988 and 1989.

» *Afterword*

This book has been a dialogue among the major players in the telecommunications industry. Twenty-two distinguished contributors from the important disciplines and institutions that impact this industry have addressed the themes of technology, regulation, management, and competition.

Although the evaluations of the economics of the industry—markets, technology, competitive forces—and the resulting prognostications clearly differ, the perspective of each contributor has validity. Underlying one or more of these perspectives is some "truth" that will unfold as a result of the interaction of economics, the dynamics of technology, and market pressures.

These forces cannot be easily bounded. The history of telecommunications is witness to that. The outcomes of many policy initiatives, technologies, and competitive entries were not predicted easily—if at all.

Also, the exogenous forces of the telecommunications customers, firms, courts, legislators, and regulators will all have their impact. The outcome will be determined by the interplay of these forces.

Whatever the result, each participant must account for, and in some way accommodate, the issues raised here. The solutions that do not "fit" these issues will fail. The solutions most likely to succeed will be those most nearly providing "win-win" answers to most of the contributors' points of contention.

One thing is clear: the forces at play are strong and in motion today.

We trust that this set of readings has given the reader a fair overview of these key concerns. The editors leave it to the reader to judge the future of the industry. We only hope that the readings have provided the necessary raw material out of which the more insightful will fashion a useful vision of the future of this critical industry.

>> Index

» *About the Editors*

James H. Alleman joined the Policy Research Division in the Office of Telecommunications as a presidential intern. He remained with the Department of Commerce until 1978, when he joined GTE Service Corporation, where he held positions of increasing responsibility in the policy research area. In 1985, Dr. Alleman took a leave of absence from GTE to accept a two-year position with the International Telecommunications Union as a telecommunications economist.

He is currently Director of the International Center for Telecommunications Management and Associate Professor, College of Business, University of Nebraska at Omaha. He also is chairman of the International Telecommunications Society.

Dr. Alleman has been responsible for local measured service research and a variety of cost, pricing, and policy analyses and initiatives. He has numerous publications dealing with telecommunications in these areas, including *The Pricing of Telecommunications Service*, published by the ITU. He was instrumental in founding an economic research group at GTE Laboratories, and has developed and organized several conferences and seminars, including the Economic Research Seminars for Africom '86 in Narobi and Telecom '87 in Geneva.

Dr. Alleman received his A.B. and M.A. degrees from Indiana University and his Ph.D. in economics from the University of Colorado.

Richard Emmerson is president of both INDETEC Corporation, which provides consulting, modeling, and related services pertaining to information decision technologies, and Emmerson Enterprises, Inc., which is a vehicle for his seminar work. He occasionally teaches at University of California, San Diego, where he was an Assistant Professor of Economics and lecturer from 1971 to 1980.

Dr. Emmerson specializes in managerial economics, the economics of law and regulation, and the use of economic and financial information for competitive decision making, specifically in the telecommunications industry. He teaches cost and profitability measurement courses for Bellcore TEC, USTA, AT&T, and others in the telecommunications industry.

Dr. Emmerson has authored dozens of professional articles based on his research in mathematical economics and regional economic modeling; he is currently supervising a major National Telecommunications Demand Study.

Dr. Emmerson received his B.A. and M.A. degrees in economics from Humboldt State University and his Ph.D. from the University of California, Santa Barbara.

» *Contributing Authors*

Loretta Anania
Research Associate
Research Program on Communications Policy
Massachusetts Institute of Technology
Cambridge, Massachusetts

Thomas Chema
Chairman
Ohio Public Utilities Commission
Columbus, Ohio

Ron Choura
Supervisor—Regulatory Affairs Section
Communications Division
Michigan Public Service Commission
Lansing, Michigan

A. Gray Collins, Jr.
Vice President—Revenues and External Affairs
C&P Telephone Companies
Washington, D.C.

Peter Cowhey, Ph.D.
Associate Professor of Political Science
Graduate School of International Relations and Pacific Studies
University of California at San Diego

Larry F. Darby, Ph.D.
Principal
Darby Associates
Washington, D.C.

Thomas A. Gajeski
Director—Technical Liaison
United States Telephone Association
Washington, D.C.

Michael L. Goetz, Ph.D.
President
PNR & Associates, Inc.
Jenkintown, Pennsylvania

Carl E. Hunt, Jr., Ph.D.
Chief Economist
Colorado Public Utilities Commission
Denver, Colorado

John M. Jensik
Director—Regulatory and Government Policy
GTE Service Corporation
Irving, Texas

Bridger M. Mitchell, Ph.D.
Senior Economist
The RAND Corporation
Santa Monica, California

Sharon L. Nelson
Chairperson
Washington Utilities and Transportation Commission
Olympia, Washington

Eli M. Noam, Ph.D.
Commissioner
New York State Public Service Commission
New York, New York
Professor (on leave)
Columbia University Graduate School of Business
New York, New York

Anthony G. Oettinger, Ph.D.
Professor and Chairperson
Program on Information Resources Policy
Center for Information Policy Research
Harvard University
Cambridge, Massachusetts

Alan Pearce, Ph.D.
President
Information Age Economics
Bethesda, Maryland

Thomas E. Quaintance
Director—Rate Development and Cost Analysis
National Exchange Carrier Association, Inc.
Whippany, New Jersey

Kenneth R. Raymond
Staff Director—Technical Regulatory Analysis
NYNEX Corporation
White Plains, New York

Leland W. Schmidt
Vice President—Industry Affairs
GTE Service Corporation
Stamford, Connecticut

Richard Jay Solomon
Consultant
Media Laboratory
Massachusetts Institute of Technology
Cambridge, Massachusetts

William E. Taylor, Ph.D.
Vice President
NERA
Cambridge, Massachusetts

John T. Wenders
Professor
Department of Economics
University of Idaho
Moscow, Idaho

Bruce J. Weston
Telecommunications Counsel
Office of the Consumers' Counsel
State of Ohio
Columbus, Ohio

Stuart Whitaker
Vice President
Information Age Economics
Vienna, Virginia

John W. Wilson, Ph.D.
President
John W. Wilson Associates
Washington, D.C.